Risk and Systems

Risk is related to the magnitude and uncertainty of an output (consequence or outcome); outputs take on different identities in different disciplines and situations. Risk is peculiar to each stakeholder and the measurement scale for risk depends on the stakeholder's value system. Risk management provides a way of addressing the issues associated with the magnitude and uncertainty of outputs. This book provides a distinctively rational treatment of risk and risk management, based on a systems approach. The book's treatment applies to all disciplines and sets out the principles of risk and risk management as well as looking at a range of applications and more specialist tools and approaches. The book:

- Develops a risk framework through a systems approach
- Offers a challenging and fresh approach for infrastructure engineering, construction and project management in general

The book will suit students and practitioners alike.

Risk and Systems
With Applications in Infrastructure Project Management

David G. Carmichael

CRC Press
Taylor & Francis Group
Boca Raton London New York

CRC Press is an imprint of the
Taylor & Francis Group, an **informa** business

Cover image: Shutterstock

First edition published 2023
by CRC Press
6000 Broken Sound Parkway NW, Suite 300, Boca Raton, FL 33487-2742

and by CRC Press
4 Park Square, Milton Park, Abingdon, Oxon, OX14 4RN

CRC Press is an imprint of Taylor & Francis Group, LLC

Library of Congress Cataloging-in-Publication Data
Names: Carmichael, D. G., 1948- author.
Title: Risk and systems : with applications in infrastructure project
management / David G. Carmichael, School of Civil and Environmental
Engineering, The University of New South Wales, Sydney 2052 Australia.
Description: First edition. I Boca Raton : CRC Press, 2023. I Includes
bibliographical references and index.
Identifiers: LCCN 2022038170 I ISBN 9781032381213 (pbk) I ISBN
9781032381220 (hbk) I ISBN 9781003343592 (ebk)
Subjects: LCSH: System failures--Prevention. I Risk management. I Risk
assessment. I Project management.
Classification: LCC TA169.5 .C378 2023 I DDC 620/.00452--dc23/eng/20221011
LC record available at https://lccn.loc.gov/2022038170

ISBN: 978-1-032-38122-0 (hbk)
ISBN: 978-1-032-38121-3 (pbk)
ISBN: 978-1-003-34359-2 (ebk)

DOI: 10.1201/9781003343592

Typeset in Sabon
by SPi Technologies India Pvt Ltd (Straive)

To Maria

Contents

Preface

Risk is related to the magnitude and uncertainty of an output (consequence, outcome). Outputs take on different identities in different disciplines and situations. The measurement scale particulars for risk follow from the stakeholder's value system. That is, risk is peculiar to each stakeholder. Risk management provides a way of addressing the stakeholder's issues associated with the magnitude and uncertainty of outputs.

Risk management (RM) is practised by everyone in both their personal and working lives. Personal risk management tends to be informal, while risk management in the workplace tends to be formal or semi-formal. Informal and formal risk management follow the same steps but to different levels of structure, accuracy and rigour. However, few people really understand the underlying basis of risk management; in many cases, workplace risk management is carried out unthinkingly by numbers or by rote. There is also some disarray present in the practice of risk management brought about by not only this lack of understanding but also by non-agreement on terminology. This may not be a serious issue when it comes to personal risk but, in the workplace, there are commercial, legal, safety and other imperatives which need to be argued rationally and transparently.

The book gives a cohesive, supportable and internally consistent treatment of risk and risk management. Explanations are given of the 'why' behind the elements of risk management. The book shows that many current practices relating to risk management are deficient and what needs to be changed in order to make any risk management rigorous.

Risk management, when dissected, can be demonstrated to be but a special case of systems engineering, or equivalently systematic problem solving or systems synthesis. (The term 'synthesis', in this book, is used in a problem-solving sense – Carmichael, 2013a.) Accordingly, an understanding of systems thinking sheds light on the correct way to look at risk management. The book explains the necessary background to systems thinking, and reframes risk management in systems terms. This permits consistent meanings for risk management terminology and a structured logical approach to risk management. This systems view of risk management is the book's strength, and allows the state of the art in risk management to be advanced.

Some views given in this book may be considered left field, and some might be considered controversial or maverick. Those who do not believe that the status quo of risk management is flawed will reject the ideas presented. Those who are content with the status quo may get upset; it is acknowledged that a large industry (consultants, educators, professional bodies, software suppliers, contractors, ...) exists selling its goods based on the status quo, and will not want to change. Whichever way people think, healthy discussion should be considered beneficial to the advancement of the state of the art of risk and risk management. It is hoped that the book sparks discussion. The book is intended to create in the reader a healthy scepticism of the current practices in risk management. The challenge is for people to acknowledge that the state-of-the-art needs improving and not dogmatically defend the current situation.

The book will be of interest to anyone trying to implement risk management in any context, and people trying to converse with others on risk.

The book alerts the reader to the terminology and methodology of risk management and gives readers a basis for undertaking their own risk studies and understanding others' risk studies.

There is much written on risk management, including international standards, guidelines of professional bodies, texts and technical and non-technical papers. However, when reading the risk literature, it is wise to be cautious. Many published risk assessments are very poor when judged as documents developed as aids to informed decision-making. Documents are often prepared by people not trained in the fundamentals required to understand risk fully. To fully understand risk, as demonstrated in this book, knowledge of statistics and probability is necessary, along with problem solving or equivalently systems thinking.

READERSHIP

The book has an infrastructure project management slant, but the framework presented in the book applies to any discipline.

About the Author

Professor David G. Carmichael is Emeritus Professor of Civil Engineering at the University of New South Wales, Distinguished Adjunct Professor with the Asian Institute of Technology, Fellow of the Royal Society of New South Wales, Fellow of the Institution of Engineers Australia and Life Member of the American Society of Civil Engineers. He is a graduate of the University of Sydney (BE, MEngSc) and the University of Canterbury (PhD).

Professor Carmichael is highly respected internationally as a conceptual thinker. He is regarded as a systems person, but more particularly regarded as a systems thinker coming from the orientation that engineering has given to the systems body of knowledge. Over his career, he has contributed significantly to fundamental systems thinking on engineering practices, particularly the many aspects of engineering management. His strength is in being able to take something, which has heretofore been presented in a complicated, esoteric, mystical and often ad hoc way, and simplify it, and give it structure and rationale.

Professor Carmichael has an extensive publication record with over 200 publications throughout his career, most peer reviewed internationally, and this includes 17 authored books across multiple disciplines. His sole-authored research monographs not only span multiple disciplines but they are particularly leading edge, innovative and beyond existing knowledge. Professor Carmichael has published in all the premier international journals in his discipline areas. His work is well-received internationally.

Specific significant seminal contributions of Professor Carmichael to knowledge span multiple disciplines: adaptability, flexibility and convertibility; future proofing; risk and risk management; investment; options; problem solving; planning; management and project management; management fads; construction, quarrying and surface mining emissions; contractor payments; project delivery; contracts; construction, quarrying and surface mining operations – production and cost; optimum structural design; material and structure characterisation; systems modelling; systems fundamental configurations; organisations; bias and decision making; design, constructability, work study; and civil engineering systems body of knowledge.

His breadth of technical knowledge is matched by few people. Much of what Professor Carmichael writes is considered to be left-field, and some might be considered controversial, maverick and uncompromising. Commonly, he shows that the status quo is flawed in an attempt to promote healthy discussion for the advancement of the state of the art of the professions. The challenge for Professor Carmichael has always been to get people to acknowledge that the state-of-the-art needs improving and not dogmatically defend current situations.

Significant sole-authored books by Professor Carmichael include:

- Future-proofing – Valuing Adaptability, Flexibility, Convertibility and Options, Springer Nature, Berlin, Singapore, 2020.
- Infrastructure Investment: An Engineering Perspective, CRC Press, Taylor and Francis, London, UK, 2014.
- Problem Solving for Engineers, CRC Press, Taylor and Francis, London, UK, 2013.
- Project Planning, and Control, Taylor and Francis, London, UK, 2006.
- Project Management Framework, A. A. Balkema, Rotterdam, The Netherlands, 2004.
- Disputes and International Projects, A. A. Balkema, Rotterdam, The Netherlands, 2002.
- Contracts and International Project Management, A. A. Balkema, Rotterdam, the Netherlands, 2000.
- Engineering Queues in Construction and Mining, Ellis Horwood Ltd. (John Wiley and Sons Ltd), Chichester, UK, 1987.
- Structural Modelling and Optimization, Ellis Horwood Ltd. (John Wiley and Sons), Chichester, UK, 1981.

Part A

Development

Chapter 1

Introduction

1.1 BACKGROUND

Risk is related to the magnitude and uncertainty of an output (consequence, outcome). Outputs take on different identities in different disciplines and situations. The measurement scale particulars for risk follow from the stakeholder's value system. That is, risk is peculiar to each stakeholder. Risk management provides a way of addressing the stakeholder's issues associated with the magnitude and uncertainty of outputs.

Risk management (RM) has been a formal part of the practice of technical disciplines, engineering, project management, factory management, business management, corporate management and other workplace endeavours for a number of decades now, although it has been practised in other disguises and without formalism for much longer. The move to a more formal recognition of risk and the need for risk management appears to have been strongest: in projects involving large capital with potentially large losses, particularly in the defence, oil and gas, aerospace and civil engineering sectors; in the insurance industry; in reliability; in accident and safety matters; in environmental and natural resource considerations; in health matters such as diseases and pandemics; in investment; in daring human projects into the unknowns of space, the oceans and remote landscapes; and where extreme natural hazards such as floods, tsunamis, cyclones/hurricanes/typhoons, tornadoes and earthquakes exist. But most management today incorporates some recognition of risk management, reflecting the importance of understanding risk in all situations.

Legislation now exists with a risk framework (though from a rigorous risk management viewpoint, the current legislation is seen to suffer from confusion as to what risk actually is). In the workplace, there may be a legislated requirement for recognition, evaluation and response related to risks. This is particularly so in workplace health and safety, dealing with workplace accidents resulting in injury or even death. It is also so in the environmental area where the impact of any development or change has to be assessed and aired for public comment. Specific legal obligations may be attached to these. Professionals practise risk management as a way of reducing their liability, or colloquially 'to cover their derrieres'. Rather than relying on good luck or favourable conditions, risk management practices are adopted.

When people talk of 'contingency plans', 'worst case scenarios' or 'what-if' analyses, they are using terminology shared by risk management. Experience and 'lessons learned' is fed back to future risk management practices.

Much of people's daily lives, at home and at work, is influenced by uncertainty. Decisions are made and work is carried out in uncertain circumstances. Decisions are dependent on events which may be only partly influenced by, or unable to be influenced by, the practician. Uncertainty, here goes hand in hand with probability, likelihood or frequency of occurrence (all used with similar intent), in contrast to determinism (Carmichael, 2014, 2016a).

This uncertainty (leading to the risk associated with this uncertainty) is widely present. When investing money, building something, administering a contract etc., uncertainty is present. Risk only exists in the presence of uncertainty. With certainty, that is determinism, risk does not exist.

Risk management is commonly broken into a series of understandable steps, with perhaps iterations, beginning with a *Definition and Context* step, followed by a *Risk Source Identification* step, *Analysis and Evaluation* step and ending with a *Response* step. Ongoing monitoring and review may ensue.

The origins of present-day risk management are not clear, though some writers suggest that it stems from the insurance industry. In a less formal disguise, risk management has always existed. It is also unclear when the term 'risk management' was first used in its present-day sense, when the term first became fashionable, and what the reason was for adopting the terminology. After reading this book, the reader might be able to offer a more appropriate term to use.

The techniques of risk management are very broad. The techniques used are borrowed from everywhere, for example, problem solving and most analysis, according to how useful the technique is for the situation at hand. The applications of risk management are likewise very broad and, in principle, risk management has relevance to any situation – workplace or otherwise.

Within industry, there are many people who profess to offer risk management services. Interestingly, many cannot give a consistent definition of risk, and many are ad hoc in their approach. The results of a risk study might be presented as something magical, but this book shows that the underlying methodology is straightforward.

Some people, so enchanted with risk, believe risk management subsumes all other management functions. It is put forward as a universal approach to all situations. The reality is that risk management grabs bits and pieces from everywhere and pretends to be all-subsuming. After reading this book, the reader should be able to establish a viewpoint on whether risk management has no content of its own, risk management subsumes all else, or something intermediate between these two extremes.

The question that might be asked is: Why risk management? The usefulness of risk management is that it forces an awareness on people to explicitly consider uncertainty. It gives a formal, rather than an ad hoc, approach to dealing with uncertainty. However, this awareness could be there independently of any risk management formalism. With such an interpretation, the body of knowledge grouped as risk management could be regarded as being superfluous and not needing to be considered as an identifiable entity.

The literature and practice surrounding risk and risk management is varied and often not internally consistent. There may be an unawareness by practitioners of this, or there is confusion in practice because of this. This is believed to be brought about by multiple existing definitions for the term risk, and practitioners not understanding the underlying systems engineering (systematic problem solving or systems synthesis) roots of risk management or even the roots of management generally (Carmichael, 2013a). Some people get it correct, but most people do not. The book expresses logically-thought-through positions on the topic. It examines what people understand by the term risk, the various alternative meanings of risk including lay and technical uses and its interpretation from a systems viewpoint. It looks at risk management, the way it is practised and its relationship to systems engineering (systematic problem solving or systems synthesis). The book's intent is to move forward the state of the art on risk in terms of definitions and practice. It is argued that systems thinking offers a vehicle for doing this.

The book shows that discussion on risk can be reduced to a few simple ideas, rather than as it is commonly presented in the literature as something complicated and mysterious. This is achieved through systems thinking. Everything about risk is embodied in its definition as developed in this book. Everything about risk management is embodied in systematic problem solving.

Risk pervades nearly everything that professionals do, and appears continually in the technical and non-technical literature and conversations. Accordingly, it would seem that it is well understood. However, on closer look, this is not the case, primarily because of its multiple definitions, because of the way risk is measured and responded to and because of the lack of explicit realization that systems synthesis is involved. Dictionary definitions of risk are not helpful in any discipline. Risk is typically viewed from a downside focus, such as a loss. However, it is also possible to consider the upside or gain. However, gains with attached uncertainty might be viewed as being unattractive or with reluctance because of the unknown implied by the uncertainty.

Generally, risk management borrows tools and techniques from other technical areas on a needs basis and, in principle, risk management should not be considered separate from other management practices. It is usual management but with uncertainty incorporated. It appears to be given its own domain in order to emphasize this uncertainty inclusion, and to ensure that uncertainty is considered in any management actions. However, provided uncertainty is acknowledged, there is no real need to even introduce the term risk, or pretend that risk management is anything special. Accordingly, some people take the view that risk management does not exist in its own right. The same people state, and quite correctly, that people have been doing risk management always and definitely long before the term 'risk management' was coined. And people still today perform risk management without expressly mentioning the term or mentioning any formalities of risk management steps. For example, a pedestrian choosing how to cross a street and avoid being injured by motor vehicles, or a swimmer avoiding sharks, crocodiles and jellyfish. And so the question might be asked: Why the term 'risk management'? It has become a very trendy and catchy term, and professing expertise in this area provides an income for many consultants, and employment for many others. It has an analogy with 'value management', 'reengineering', etc., in this regard (Carmichael, 2013a). The word 'risk', in many quarters, has become one of those overused terms that fit easily within the buzzword bingo game played at project, team or management meetings, or might be found in a Dilbert book.

Once upon a time, people were happy dealing with uncertainty, and making decisions based on uncertain outputs, and nobody saw the need to use the term 'risk', let alone the more grandiose term 'risk management'. Risk management was carried out less formally, and the term risk management was not used. Today, the use of the term risk management is all pervasive, but interestingly most people get the associated practices wrong – people sometimes get the right results, but without realizing what they are doing. The question that follows is: Why adopt the formality of risk management, if current practices largely do not understand what is happening?

The presence of uncertainty and probability may be a barrier to a large number of people eventually understanding risk. It is the author's observation that a great many people do not understand probability and feel more comfortable operating in a deterministic world. Technical people generally favour deterministic models, even though their education may include instruction on probability. Many non-technical disciplines, which have an interest in risk, have no basic educational grounding and understanding in probability and problem solving, and hence will struggle in truly understanding risk, as well as in applying risk management properly. Much of the present-day confusion surrounding risk might be because of people lulling themselves into thinking that they understand risk and risk management when they do not have the background to do so. Understanding and conducting proper risk management may be beyond lay persons and most disciplines; this points to risk management becoming the domain of technical professionals, such as engineers and like-educated specialists, much like the structural design of a building is only entrusted to qualified structural engineers.

Discussing the costs and benefits of undertaking risk management has no point because risk management, or its equivalent without the name, is the appropriate approach that takes into account uncertainty – it is systematic problem solving. It is the approach which should be adopted in order to fully understand and manage the risk sources (inputs), the conversion of inputs into outputs and the outputs themselves.

1.2 BOOK OUTLINE

The book is presented in two parts. The development of an understanding of risk and risk management comprises Part A, while applications follow in Part B. The book provides an original and definitive treatment of risk and risk management. Risk management is examined in terms of standard, accepted steps, but explained in terms of the underlying fundamentals. All risk is treated in a common fashion. Risk management is considered from the viewpoint of the stakeholder for whom any risk study is being undertaken. A wide range of practices engaging with risk management are covered.

The book outlines what risk is if systems thinking is adopted. In the light of this, it examines the multiple published and unpublished definitions that exist for risk and gives comment on these. Risk management is examined in terms of standard, accepted steps, and comment is given on the practice of risk management. The book does not deal with the over-arching organizational issues associated with risk management or psychological issues involving people's value systems. No distinction is made in the book between categories of risk such as pure/speculative or systemic/non-systemic, or between different disciplines and applications that deal with risk. All risk is treated in a common fashion. No distinction is made between organizations or people or disciplines when referring to risk management; risk management is considered from the viewpoint of the stakeholder for whom the risk study is being undertaken. In the following, stakeholder (singular) is used, but the commentary can also apply to multiple stakeholders. A stakeholder is a person or organization for whom the risk study is being performed. The person conducting the risk study on behalf of a stakeholder is referred to as a practician here, where a practician may be an engineer, manager or other stakeholder agent; when conducting a risk study on behalf of himself/herself, a practician is also the stakeholder.

The book provides a truly rational treatment of risk and risk management, which applies across all disciplines, technical and non-technical. All risk is treated in a common fashion. The underlying basis for the book is a systems approach.

Because the risk literature is very large, representative publications only have been selected for comment, it not being possible to comment on everything written on risk.

1.3 TERMINOLOGY

The literature is seen to use multiple terms to mean the same thing. This book uses terms as follows.

Systems approach: Equivalent terms are systems engineering (in the sense of Hall, 1962), systematic problem solving (in the sense of Carmichael, 2013a), systems synthesis (in the sense of Carmichael, 2013a), systems thinking (in the sense of Carmichael, 2013a) or simply systems.

Inputs and outputs: A reference to the plural system inputs and system outputs may include reference to the singular input and singular output. (Singular here refers to grammar, not a mathematical meaning.)

Inputs: Equivalent terms are risk sources, risk factors or risk events (but not in the sense of safety factors, factor analysis, events in a sample space or event trees). Inputs: (i) may be able to be influenced by the practician; or (ii) may be beyond the influence of the practician.

Outputs: Equivalent terms are consequences or outcomes.

Category: Category describes the type of risk, for example, a risk related to health issues is put in the category labelled health risk, while a risk related to finances is put in the category labelled financial risk. This is for compartmentalization convenience only, and does not signal a different treatment of risk.

Evaluation: An equivalent term is assessment.

Response: Equivalent terms are control or treatment.

Practician: A practician is a person conducting a risk study on behalf of a stakeholder. A practician may be an engineer, manager or other stakeholder agent. When conducting a risk study on behalf of him/herself, a practician is also the stakeholder.

Practitioner: A practitioner is someone working in industry.

Stakeholder: In the book, stakeholder (singular) is used, but the commentary can also apply to multiple stakeholders. A stakeholder is a person or organization for whom the risk study is being performed (by a practician).

Value system: The stakeholder's value system may be influenced by public opinion, psychological issues, perceptions, legal requirements and background, among other matters.

Expected value: The expected value of something equals magnitudes weighted with probabilities of occurrence, or the combined product of magnitudes and probabilities. With a probability distribution, expected value is equivalent to the mean or average, and is a measure of central tendency (Benjamin and Cornell, 1970; Ang and Tang, 1984); expected value, average and mean are precisely defined within treatises on probability and statistics.

Expected utility: Expected utility of something equals utilities weighted with probabilities of occurrence, or the combined product of utilities and probabilities (Benjamin and Cornell, 1970; Ang and Tang, 1984). (The economics use of the term utility is not used.)

Uncertainty: Uncertainty implies variability or variation. It goes hand in hand with probability, likelihood or frequency of occurrence (all used with similar intent), in contrast to determinism (Carmichael, 2014, 2016a). The associated variables and models are probabilistic. (The decision theoretic view of uncertainty implying an absence of probabilities is not used.)

Possibility: Possibility is used in the sense of a sample space (where probabilities are attached), that is a value in a collection of many conceivable values. The values could relate to inputs, outputs or other matter depending on the application.

Probability: An equivalent term is chance. It indicates a relative likelihood, frequency or weighting (all used with similar intent) attached to a value in the sample space. Probabilities, chances and likelihoods may derive from some objective root, for example, data, or be subjectively established, and can arise from any suitable origin.

Determinism: Equivalent terms are certainty or no uncertainty. Risk only exists in a non-deterministic (probabilistic) situation; something is regarded as 'risk-free' when a certain (equivalently, no uncertainty) output exists.

Dispersion: With a probability distribution, measures of dispersion include variance, standard deviation, coefficient of variation and range. Variance, standard deviation, coefficient of variation and range are precisely defined within treatises on probability and statistics.

Chapter 2

Some fundamental systems ideas

2.1 OUTLINE

Understanding risk and risk management can best be obtained by adopting systems thinking. It is not necessary to know everything about systems ideas, but the few necessary concepts are visited here. Everything about risk management is embodied in systems engineering (systematic problem solving or systems synthesis).

Carmichael (2013a) notes, among other matters, that a system is composed of interacting or interrelated entities, elements, objects, parts or components (subsystems). Whatever does not belong to a system is referred to as the environment (and this should not be confused with the natural environment). This automatically establishes the system boundary (between the system and the environment). External to the system, inputs and outputs take place. The nature of the inputs and outputs and their characterization, depend on the system study being undertaken. In this sense, a system may also be thought of as an input-output transformation. In studying any system, some system properties or attributes might be selected as relevant to the study and other properties may be ignored. It will depend on the intent of the study.

Of the various systems schools, perhaps the most straightforward and general systems approach is that based on control systems theory (Carmichael, 2013b). It is compatible with the systems engineering approach of Hall (1962). Background to this systems framework is given below. This book's discussion centres on a single-level system for explanation purposes, but subsystems and lower levels could be introduced for more completeness. (A subsystem is a system.)

Risk management can be shown to be a special case of systems engineering (in the sense of Hall, 1962) or equivalently systematic problem solving or systems synthesis (in the sense of Carmichael, 2013a,b), and shares common system structures with design, decision making, planning and the other forms of management (Carmichael, 2013a). As such, objective function(s), constraints and a system model are present and need elucidation (preferably

DOI: 10.1201/9781003343592-3

in quantitative forms for definiteness, but not necessarily), as in any synthesis. This is developed further below.

This book reasons that the only acceptable meaning for risk is that it is related to the magnitude and uncertainty of an output (consequence or outcome). Outputs take on different identities in different disciplines and situations. That is, risk is a function of output, Risk = f(Output). Output is characterized by its magnitude and likelihood (frequency or probability – all used with similar intent – as a result of uncertainty). The measurement scale particulars for risk follow from each stakeholder's value system. That is, the pairs {output magnitude; output likelihood} convert to risk values peculiar to each stakeholder. The output contains uncertainty; with certainty, there is no risk. This is the only meaning for risk that provides a useable and consistent approach to risk management. (The use of the term 'function' is not intended to imply a quantitative-only approach. A solely mathematical approach is not intended. The book allows nominal, ordinal, interval and ratio scale types.)

Outputs result from inputs (the risk literature uses risk sources, and several other terms) – some with uncertainty, some without.

That is, as a precursor to establishing risk, pairs of output information are required – {output magnitude; output likelihood}. Here likelihood or probability may be the probability associated with an output or an output exceedance (or non-exceedance) probability, noting that output magnitude may have a sample space. The full probability distribution for an output might not be considered because of the difficulty of incorporating this in any analysis.

Particular applications might report output in ways other than as a pair, {output magnitude; output likelihood}, including:

- Either output magnitude or output likelihood is reported for a constant value of the other.
- Expected value of the output, for so-called 'risk-neutral' attitudes.
- Expected utility of the output, where the utility measure is that of the stakeholder for whom the risk study is being carried out – different stakeholders will have different utility functions or curves.

The second and third ways of reporting require the output to follow an interval or ratio scale type (Section 2.3 and Carmichael, 2013a).

Output magnitude might be defined relative to some desired or base level, rather than an absolute magnitude. Uncertainty goes hand in hand with probability, likelihood or frequency of occurrence (all used with similar intent), in contrast to determinism (Carmichael, 2014). The variables and models being used are probabilistic. (Initiating) inputs such as extreme weather and human and equipment functioning originate the uncertainty. Risk is related to the output (containing uncertainty) of the system being examined, resulting from inputs (some with uncertainty, some without); in general, a system has multiple inputs and multiple outputs (and can be mul-tilevel, with interacting subsystems).

2.2 SYSTEMS BACKGROUND

The following gives a broad outline of the systems framework referred to in this book. The outline extracts from control systems theory (Carmichael, 1981, 2006, 2020c, 2021c,d), and systems engineering in the sense of Hall (1962) or the equivalent in systematic problem solving or systems synthesis (Carmichael, 2013a,b).

2.2.1 Variables

In any application, the main and relevant system variables are those of input, state and output:

Output variables are variables reflecting the external, measurable or observable behaviour (performance) of the system. (Note, that the systems behavioural term 'response' is avoided here so as to not confuse with the risk management *Response* step.)

State variables are variables reflecting the internal behaviour of the system. The system's state may be non-observable and non-measurable. In many cases, the state and the output turn out to be, or are assumed to be, the same – there is a one-to-one relationship between output and state.

Input variables are of two types:

- *Control* (decision, action, design) variables which can be manipulated, influenced or chosen by the practician in order to drive the system behaviour (output, performance) in a desired way, usually according to some objective function(s). (The term control is <u>not</u> used here in the lay sense or popular management sense of containment.) Control variable values are regarded as input to the system by the practician.
- Those over which the practician has no influence, or chooses to have no influence. These are not given any special name.

Some inputs might be considered to have no uncertainty attached to them, while others might have uncertainty attached to them. Systems are com-posed of interacting subsystems, and these system variables are present at all levels in a system hierarchy. A subsystem is a system.

2.2.2 Fundamental configurations

It is possible to make a distinction among what professionals, and people generally, do (Carmichael, 2013a). The way people approach their trade or work may be categorized or shown to fit three types of configurations, of which two are most relevant to risk management, referred to below as analysis and synthesis.

Consider a system represented by Figure 2.1a. Using the previously made distinction between the variables, let the control inputs be A, the model of the system B and the outputs C, as in Figure 2.1b. For other inputs, known or estimated, the fundamental configurations of analysis and synthesis become: Analysis – given A and B, obtain C; Synthesis – given B and C, obtain A. (The third configuration – investigation – is given A and C, obtain B.)

Risk management, like all management, belongs to synthesis (Carmichael, 2013a). In general, a result in synthesis is non-unique, whereas the result in analysis is unique. Non-uniqueness in the synthesis configuration might be removed through additionally including some optimality measure(s) (here referred to as an objective function or its plural), and/or results reduced through including constraints. (Something is only optimum or best with respect to an objective function. Without an objective function, the terms optimum, maximum, minimum or best have no meaning.)

Any general synthesis contains three components: a system model, an objective function(s) and constraints (Carmichael 2013a). And so risk management must recognize the presence of these components. Objective function(s) and constraints are selected from the point of view of the stakeholder, for whom the risk study is being undertaken. Different stakeholders will select different objective function(s) and constraints, and hence will have different results in their risk management. The model provides the input-output transformation needed in analysis.

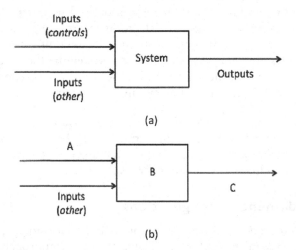

Figure 2.1 System representation: (a) variables described; (b) using notation.

People tend to adopt a pragmatic approach to tackling synthesis through a simple path of iterative analysis (rather than a direct optimization approach), where typically the steps of:

- *Definition.*
- *Objective function(s) and constraints statement.*
- *Alternatives generation.*
- *Analysis and evaluation.*
- *Selection.*

are adopted, along with feedback and iteration between steps (Hall, 1962; Carmichael, 2013a,b). (Figure 2.2.) (The system may also be adjusted in an

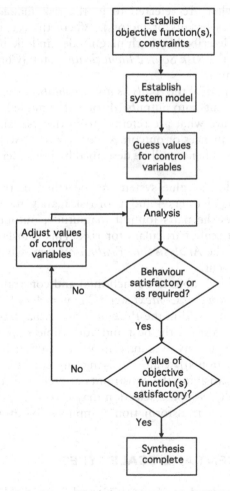

Figure 2.2 Iterative-analysis version of synthesis (definitional step not shown) (Carmichael, 2013a); non-adaptive.

adaptive control sense – Carmichael, 2015.) The iterations are not inherent in synthesis but occur because of the analysis-based mode of attack. Risk management, being a form of synthesis, is typically tackled this way. This means that risk management is a special case of the more general systems engineering, systems synthesis or systematic problem solving.

The remainder of this book uses the terminology and systems ideas just covered.

2.2.3 Applied to risk management

To apply the above systems ideas to risk management, the following thinking is required:

- Inputs are what are referred to in the risk management literature as risk sources (events or factors). Where these contain uncertainty they will be described by both magnitudes and likelihoods. These are unearthed in the *Risk Source Identification* step (Chapters 4 and 6) of risk management.
- The analysis part of the *Analysis and Evaluation* step (Chapters 4 and 7) converts inputs into outputs (through the model).
- The outputs are what are referred to in the risk management literature as consequences or outcomes. For risk to exist, these will contain uncertainty and hence will be described by both their magnitudes and likelihoods.
- The stakeholder's value system is unearthed in the *Definition and Context* step (Chapters 4 and 5) of risk management. This value system translates the pairs {output magnitude; output likelihood} into measurement scale particulars for risk. This translation is the evaluation part of the *Analysis and Evaluation* step (Chapters 4 and 7) of risk management.
- The stakeholder's objective function(s) and constraints are unearthed in the *Definition and Context* step (Chapters 4 and 5) of risk management. These are used in the *Response* step (Chapters 4 and 8) of risk management to select the best and admissible response (adjustment), via synthesis. This involves maximizing or minimizing something (the objective function(s)) while satisfying any constraints. Responses or adjustments satisfying the constraints are referred to as being admissible. Generally, response selection may involve something to do with the inputs, or the transformation of inputs into outputs.

2.3 MEASUREMENT AND SCALE TYPES

The reader is referred to Lehmann (1979) and Carmichael (2013a) for more detailed comment on measurement and scale types.

In risk management, measurements and scale types are central. The construct may be a belief, quantity or other. People might be more at ease working with quantitative scales for these constructs, although the use of quantitative scales in some situations may not be appropriate. Quantitative scales permit reduction and graphical display, and also permit the use of computers.

Scale types may be classified as (Figure 2.3):

- Nominal.
- Ordinal.
- Interval.
- Ratio.

The choice of scale type may be dictated by the construct being measured. The scale type influences the form any analysis and evaluation takes. The first two scale types in Figure 2.3 are nonmetric; the last two are metric. They are listed in order of the number and types of calculations that can be performed on measurements done according to the scale types; nominal scale types permit only a few calculations while ratio scale types permit all statistical calculations.

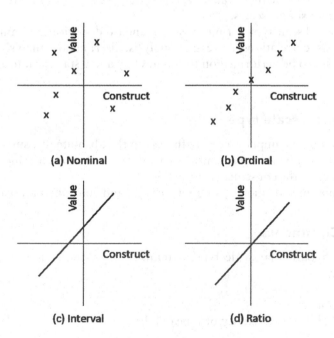

Figure 2.3 Example scale types (after Lehmann, 1979).

2.3.1 Nominal scale type

A nominal, cardinal or categorical scale type has no meaning in itself and does not relate to other scale types. There is no relationship between the amount of the construct and the numerical measure (Figure 2.3a). It denotes quantity but not order in a group.

Examples include colours allocated numbers, and numbered responses to multiple choice questions.

2.3.2 Ordinal scale type

In an ordinal scale type, there is a relationship between the amount of the construct and the numerical measure but not the absolute value of the measure (Figure 2.3b). It denotes order, quality or degree in a group such as first, second, ... An ordinal scale type permits medians and percentiles to be calculated. The intervals between values cannot be related.

Examples include rankings, paired comparisons and semantic scales.

2.3.3 Interval scale type

Where the differences between numerical measures, rather than the absolute values of the measures, are important, then this is an interval scale type (Figure 2.3c).

Examples include thermometers, bipolar adjective scales (discrete or continuous) and survey scales.

An interval scale type permits means, standard deviations, parametric statistical tests, correlations, regression analysis, discriminant analysis and factor analysis, to be performed on the data. Only a few statistical tools cannot be used.

2.3.4 Ratio scale type

A ratio scale type implies meaningfulness to the absolute measurement values and the intervals between measurement values. As well, a value 0 implies the absence of the construct (Figure 2.3d).

Examples include money, amount surveys, and constant sum scales.

2.3.5 Comments

In using a particular scale type, a number of matters raise their heads, including:

- Validity.
- Reliability, consistency or repeatability.

- Bias.
- Efficiency.
- Errors.

Validity refers to the ability of a measure to represent a construct. Reliability is in the sense of repeatability. Bias occurs when a result is obtained different to the true value. Efficiency is in terms of economy of effort to obtain a result.

Chapter 3

What is risk?

3.1 ACCEPTABLE DEFINITION

The book reasons that risk is related to the output of a system: Risk = f(Output). An output is characterized by its magnitude and likelihood. The measurement scale particulars for risk follow from each stakeholder's value system. That is, the pairs {output magnitude; output likelihood} convert to risk values peculiar to each stakeholder. The output contains uncertainty; with certainty, there is no risk. This is the only meaning that provides a useable and consistent approach to risk management. (The use of the term 'function' is not intended to imply a quantitative-only approach. A solely mathematical approach is not intended. The book allows nominal, ordinal, interval and ratio scale types – Chapter 2.)

Outputs here refer to consequences or outcomes. The outputs result from inputs (the risk literature uses risk sources, events or factors). Some inputs have uncertainty, some do not. Uncertainty here goes hand in hand with probability, likelihood or frequency of occurrence (all used with similar intent), in contrast to determinism (Carmichael, 2014).

An output may be a single value or a random variable. As a random variable, output may have a continuous or discrete sample space. (More complicated descriptions for output appear not to be considered by practitioners and are not covered here. Everything in practice appears to be converted to these two cases – single value or random variable.)

The uncertainty associated with a single value may be expressed as a likelihood, probability or frequency of occurrence (all used with similar intent). The uncertainty associated with a random variable may be represented by a probability distribution – probability density function, probability mass function or cumulative distribution function – from which probabilities or probabilities of exceedance (or non-exceedance), based on that output's sample space, may be obtained.

It follows that as a precursor to establishing risk, pairs of output information are required – {output magnitude; output likelihood}. Here likelihood may be the probability associated with an output or an output exceedance (or non-exceedance) probability. The full probability distribution for the

DOI: 10.1201/9781003343592-4

output might not be considered because of the difficulty of incorporating this into an analysis.

Particular applications might report output in ways other than as pairs of {output magnitude; output likelihood}:

- Either output magnitude or output likelihood is reported for a constant value of the other.
- Expected value of the output, for so-called 'risk-neutral' attitudes.
- Expected utility of the output, where the utility measure is that of the stakeholder for whom the risk study is being carried out – different stakeholders will have different utility functions or curves.

The second and third ways of reporting require the output to follow an interval or ratio scale type (Chapter 2 and Carmichael, 2013a).

As well, output magnitude might be defined relative to some desired or base level, rather than an absolute value.

The stakeholder's value system establishes the risk measurement scale particulars that the stakeholder places on the output pairs. Different stakeholders have different value systems and this leads to different measurement scale particulars for risk and hence different risks described by stakeholders for the same output pairs.

3.2 DISARRAY

Throughout this book, the disarray that exists in the current literature and practice associated with risk and risk management is highlighted, in an attempt to promote discussion on risk and advance the state of the art. Multiple conflicting meanings and usages of the term risk can be found in the literature and in communications. Generally, these meanings and usages are not internally consistent. Be aware that many meanings are seen in the literature and used in communications. Sometimes several definitions are loosely used by the one person in the one discussion. Flip-flopping from one meaning to another seems a characteristic of many people communicating on risk.

Some of the confusion over the definition of risk arises because of publications and people using the terms risk source (risk event, risk factor) and risk interchangeably, and not appreciating the distinction. Comment is given on this in Chapter 6, *RM Step – Risk Source Identification*; it is seen that people (wrongly) use the term 'risk identification' when they are referring to 'risk source identification'. A risk source (upstream) leads to a risk (downstream); they are very different.

Some further confusion occurs with the term 'risk analysis'. Comment is given on this in Chapter 7, *RM Step – Analysis and Evaluation*. A more appropriate description would be 'converting inputs to outputs'.

Risk is a pervasive part of all actions by people. Everyone (overly) uses the term risk, in both their daily lives and work lives. It would seem on the surface that the term risk is a simple, well-understood notion. However, this is not the case. Unfortunately, people have not been able to agree on a single definition of risk, while many people are loose in their usage of the term. With time, it is hoped that people will adopt the rigorous treatment of risk given in this book in order that the discipline can progress rather than go further backwards.

In spite of the confusion in the literature and public discussion as to the definition of risk, interestingly it does not seem to bother practitioners and academics. In people's daily lives the use of correct or incorrect meanings for words does not seem to matter; people get along somehow as long as the intent of the conversation is understood. But it is unacceptable that professionals do not use terminology correctly. Some unanswered questions follow: Why is this the case? Is a rigorous definition not important? How can people talk to each other meaning different things for the same word? How can such a situation be tolerated in the workplace, where the message receiver translates the message differently to that of the sender?

3.3 EXISTING DEFINITIONS

Most treatments of risk focus only on the downsides such as loss, damage, extra cost, extra time and negative outputs, and neglect the upside or opportunity such as profit or gain. Risk is usually associated with danger and peril. However, in principle, there is no need to make a distinction between upsides and downsides. Just as things can possibly go wrong, so there may be a possibility of things going well.

In line with the possibilities of both loss and gain, a distinction may be seen in the literature as follows: where only a loss or no loss can occur (pure risk); and where not only a loss and no loss can occur but also a gain can occur (speculative risk). However, this distinction is unnecessary and only complicates thinking on risk.

On examination of the literature, it is seen that risk has many different meanings. There is no consistency of usage or unanimously accepted usage. And it is not uncommon to see more than one usage in the one document or conversation. Looseness in terminology and flip-flopping between definitions does not seem to worry some people. As well, the word risk is one of those overused terms that fit easily within the humorous buzzword bingo game played at project, team or management meetings, or you might come across in something written by Scott Adams of Dilbert fame. The term is frequently used when there are more appropriate terms available, to impress, or to disguise a speaker's or writer's lack of understanding of a situation. This has the potential to cause much confusion.

Using the term risk, it is remarked, is much like the dilemma in Alice in Wonderland:

> 'When I use a word', Humpty Dumpty said in a rather scornful tone, 'it means just what I choose it to mean, neither more nor less'.
> 'The question is', said Alice, 'whether you can make words mean so many different things'.
>
> ('Through the Looking Glass', Ch. 6, Lewis Carroll)

This book is less concerned with the historical origins of the word risk than its present-day usage. The dilemma facing the misuse of the term may have been influenced in part by its historical roots, but it has been exacerbated by the present-day potpourri of non-rigorous and conflicting usages within multiple disciplines, each acting as a silo and not looking outside its own discipline, or looking to be rigorous. Different usages of the term risk across disciplines could be workable; however, it is seen that disciplines unfortunately do not use the term risk in any internally consistent way, and that the term is equally misused across all disciplines. Modern treatments of risk and risk management appear to have evolved without any noticeable paradigm shifts along the way. This book is suggesting a paradigm shift.

The natural questions that follow are: How did such a situation come about? Why do people let this situation continue? How do people communicate on risk when everyone is interpreting risk differently? The intent of any communication on risk might be understood, but clearly no deep information can be conveyed. Clearly, there is a need to standardize on a definition in order that the state of the art on risk can progress. Without agreement on what risk means, the theory and practice of risk management will not advance. Definitive meanings of terms are the hallmark of a body of knowledge in any advanced discipline; the advances to date in engineering, for example, would not have been possible if everyone was using the same technical words but with different meanings.

A selection of the range of usages of the term risk is given here. Further examples of confusion as to what risk actually is can be seen in the risk management steps discussed in later chapters. The reader is asked to contrast the explicit or implied definitions quoted in the literature with the definition given in Section 3.1.

3.4 DICTIONARIES

Most dictionary definitions of risk are unsuitable for a technical treatment of risk. Yet, it is frequently seen that papers in technical journals, books and reports use such meanings. Dictionaries commonly give multiple meanings, often conflicting, placing the reader in the predicament as to what meaning should be applied. Dictionaries give lay meanings of words and generally are not useful when it comes to working technical definitions. No mature discipline uses dictionary meanings for terms in its body of knowledge. Example dictionary definitions are:

- *The possibility of incurring misfortune or loss.*
- *Hazard.*
- *Chance of a loss.*
- *Chance of an event.*
- *The type of such an event, such as fire or theft.*
- *The amount that an insurance company stands to lose.*
- *A person or thing that may cause an insured event to occur.*
- *The possibility of suffering harm or loss.*
- *Danger.*
- *A factor, thing, element, or course involving uncertain danger.*
- *The danger or probability of loss.*
- *The variability of returns from an investment*
- *The chance of non-payment of a debt.*
- *A probability or threat of damage, injury, liability, loss, or any other negative occurrence.*
- *Potential of losing something of value.*

This list is not exhaustive. Further meanings occur in dictionaries.

3.5 POPULAR USAGE

From observations, the following risk meanings, among others, are adopted in popular usage: *hazard; uncertain event; uncertainty; probability; chance; possibility;* and *exposure to negative events.* Rigorous technical communication on risk is not possible with such ambiguous and quite different meanings.

A definition of risk as a probability or chance is unacceptable, and this can be demonstrated through examples involving extreme events (cyclones/hurricanes/typhoons, tornadoes, tsunamis, earthquakes, ...). The extreme event is the input or risk source, with magnitude and likelihood attached. The risk (popularly but wrongly) being referred to is usually this input likelihood. The extreme event's downstream outputs follow. The literature makes no distinction between the likelihood of an input and the likelihood of an output, yet the two generally are not the same, leading to possible ambiguity as to which probability or chance is being referred to. The risk cannot be the input likelihood, chance or probability; the extreme event is destined to happen but what is needed to be managed is something to do with the downstream outputs of the extreme event. If risk is being used to refer to the output likelihood then it is deficient; if performing risk management, then only output likelihood is being managed and not output magnitude. Other literature talks strangely of 'risk probability'.

A definition of risk as a possibility is also unacceptable. From elementary probability theory, a range of possibilities will usually be present where uncertainty exists. The management of possibilities, or the size of the sample space of possibilities, is not believed what is intended, because of the lack of scope available to do this. Lay usage of the terms possibility and probability (and often to mean the same thing) are not consistent with those of probability theory.

It is also observed that while a document may use one of the above meanings, the associated measurement adopted for risk is usually inconsistent with the meaning. For example, risk might be (wrongly) thought of as a probability or chance, but be measured in terms of something about output.

Cliched expressions involving risk are many. For example: *no risk; to take or run a risk; a calculated risk; at risk; risky; riskiness; risk versus reward; risk:reward ratio; risk-return; to take unnecessary risks; to take a chance; chancy;* and *hazardous.* Brave people, entrepreneurs, and explorers are said to *take risks.* Certain people in life are said to be *good risk managers,* without enlarging on what that means. All these usages generally follow dictionary meanings. Although they should have no place in technical documents and technical discussions, these usages occur frequently in such places.

Adages and sayings on risk also abound. A few examples, popularly attributed to the cited person accompanying the quotation, follow: *Only those who will risk going too far can possibly find out how far it is possible to go.* (T. S. Eliot); *People who don't take risks generally make about two big mistakes a year. People who do take risks generally make about two big mistakes a year.* (P. Drucker); *Progress always involves risks. You can't steal second base and keep your foot on first.* (F. Wilcox); *Take risks: if you win, you will be happy; if you lose, you will be wise.* (Anonymous); *Danger can never be overcome without taking risks.* (Latin Proverb); *Women try their luck; men risk theirs.* (Oscar Wilde). Clearly, the meanings are similar to those available in dictionaries and align with popular ambiguous usage. Ambiguous meanings for words are a commonly found origin of humour, and dictionary meanings for the word risk provide the source of many chuckles.

Television and other media continuously reinforce this ambiguous, and sometimes an emotive, view of risk. For example, a recent advertisement on how alcohol affects driving skills gives: *for a human body containing 0.05 grams of alcohol per 100 ml of blood, there is double the risk of having an accident* [over having no alcohol in the blood]; *for 0.10 grams, there is 7 times the risk of having an accident; for 0.15 grams, there is 26 times the risk of having an accident.* Here, risk means the probability of an accident.

The discontinued original predecessor, AS/NZS 4360 (AS/NZS, 1995), to ISO 31000 (ISO, 2009) gave conflicting and unsupportable meanings for the term risk: *the chance of something happening that will have an impact upon objectives. It is measured in terms of consequences and likelihood.* ISO 31000 (ISO, 2009) uses *the effect of uncertainty on objectives,* which can be interpreted in different ways, and does not give a useful meaning for risk. Such a definition is difficult to work with because of a lack of definiteness. These matters were discussed in Carmichael (2016a) and are not pursued further here.

3.6 OTHER OCCURRENCES

Some parts of decision theory refer to making decisions *under conditions of risk* and *under conditions of uncertainty*. *Conditions of uncertainty* refers to outcome probabilities not being known; decisions might follow, for example, maximin, maximax and minimax regret practices. Where knowledge of the possible outcomes exists and probabilities can be assigned to these, it is referred to as decision making under *conditions of risk*. The usage of the terms *risk* and *uncertainty* here is not consistent with that used in this book.

Decision theory is well established and has developed its own unique terminology. Generally, that terminology is not consistent with that adopted in systematic problem solving; terms such as *analysis, theory, model,* along with *uncertainty* are used in different senses to rigorous systematic problem solving (Carmichael, 2013a). From personal observations, this causes much confusion in those trying to understand risk.

Other discipline applications given in later chapters highlight the broadly held confusion regarding people's understanding of risk.

3.7 SUMMARY

Risk, within a technical treatment of risk, is not a *chance*, a *probability*, a *likelihood* (qualified with respect to the fixed magnitude comment of Section 3.1), or a *possibility*. A *risk source is not a risk*. Commonly, *possibilities* and *probabilities* associated with *risk sources* are not able to be influenced. These definitions are unsuitable for a rigorous development of risk management.

To understand risk management and to apply it properly and rigorously, dictionary and lay person usages of the term risk need to be put aside. This implies that people need to be bilingual – being able to understand lay meanings in social communications, but switching to a rigorous meaning when talking technically. This happens elsewhere, for example, a structural engineer's definitions of stress and strain are different to lay usage, but a structural engineer would never contemplate using the lay meanings when talking about the state of a structure under load. Why then do people do so when talking about risk?

Unfortunately, standards and many practitioners and academics use the term risk inconsistently or incorrectly. And unfortunately, very few people seem to question the status quo.

Chapter 4

Risk management (RM)

4.1 ACCEPTED STEP APPROACH

Risk exists because of uncertainty. Uncertainty may arise from quite diverse origins – the regulatory process, natural hazards, monetary matters, unproven technology, management matters, resource availability, industrial relations issues and so on. Success in any undertaking could be anticipated to depend on being aware of this uncertainty and appropriately dealing with it, for example, through the described risk management practices in this book.

The popular approach to risk management (RM) goes through a number of steps, variously described in terms related to:

- *Definition and Context.*
- *(Risk Source) Identification.*
- *Analysis and Evaluation.*
- *Response.*

Although using different words and numbers of steps, what happens within these steps is no different to that involved in systems engineering (for example, as espoused by Hall, 1962), or as interpreted equivalently in systems synthesis via iterative analysis or systematic problem solving (Carmichael, 2013a), where the steps are given broadly as:

- *Definition.*
- *Objective function(s) and constraints statement.*
- *Alternatives generation.*
- *Analysis and evaluation.*
- *Selection.*

Systems engineering (Hall, 1962), systems synthesis and systematic problem solving involve equivalent steps because they are doing essentially the same thing in systems terms. Feedback and iterative modification, for clarification, refinement and comprehensiveness purposes, may occur within and between

DOI: 10.1201/9781003343592-5

any of the steps. Risk management, like systematic problem solving, in essence, becomes trial-and-error optimization. To start the iterations in risk management, a common approach is to assume the status quo. Adjustments are then made to appropriate inputs or the system via a feedback mechanism. Activities, such as monitoring and updating, following the above steps can occur. There may also be something equivalent to 'installing' the final adjustment, as adopted in work study (Carmichael, 2013a). For a changing situation, risk management becomes an ongoing practice over time.

This means that risk management is a special case of the more general systems engineering, systems synthesis or systematic problem solving. The extension to this argument is: the subject of risk management should not be regarded as a stand-alone discipline; and the subject of risk management could be said to have no content that it can say is its own. The questions that follow (and which have wider relevance beyond this book) are: How do such special approaches come about? Why do not people start with the existing and more generic systematic problem solving and either use this or specialize this? Is it that people have trouble starting with the general and going to the specific? Is it easier and less confronting to approach any situation from the specific? While specialization of practices is perhaps inevitable to some extent, it also has the effect of limiting people's perspective, and this is what appears to have happened in risk management, leading to the confusion observed in the terminology and practice of risk management.

4.2 ISSUES IN RISK MANAGEMENT

One issue facing risk management is that it is practised in everyday life, albeit informally and subconsciously. Few workplace disciplines can say that they share something with what is done out-of-work. People take precautions and behave related to identified risk sources (inputs). The textbook example is a pedestrian crossing a street and avoiding being hit by a vehicle. People accept 'risk' in everyday life, even in their own homes and with their own chosen acquaintances. Younger people seem to accept more than older people, what might be perceived by others as higher 'risk', or perhaps the data and analysis of younger people are more incomplete compared to those used by older people, while their value systems are different. Even though people may not think of this as risk management, it is risk management nonetheless. Although unstated, the steps people go through in an informal fashion out of the workplace are the same as those of formalized risk management practised in the workplace.

A second issue facing risk management is that there is a societal (and often a legal and commercial) expectation that nearly everyone in the workplace practises risk management. This necessarily means that risk management is often practised at a basic level, without understanding and often wrongly. Risk management becomes procedural through checklists, standard forms, ticking boxes and the use of practices which are blindly followed. Some people think that by filling in a form or completing a checklist, they are managing risk. People go through the motions, cookbook style and do not think about what they are doing. People are focussed on completing the steps rather than understanding the issue at hand. It is a means to an end for them. They do not appreciate that risk management is about systematic problem solving. Hence most people get risk management wrong. And the thinking which leads to mistakes gets perpetuated through the popular literature. People get answers, and that keeps the boss, regulators or the client happy. But they do not understand what they are doing. However, if risk management is looked at from the structure offered by systematic problem solving, answers are not only obtained, but also understanding and the avoidance of wrong answers follows. Additionally, systematic problem solving simplifies and structures thinking. Unfortunately, most people do not know what systematic problem solving is. Systematic problem solving should be included in all educational qualifications but, unfortunately, it gets squeezed out by short-term-applicability technical content (Carmichael 2021a,b,c,d).

Risk management can be done to different degrees of detail, from broad to fine. Further, within any risk management attempt, it can also be done in the broad-to-fine order sequentially, each time establishing better understanding, but involving more work.

The author's observations are that mistakes in implementing risk management are common, even with industry people who make a living out of specializing in risk management or who purport to have risk management as their expertise, primarily because people do not fully appreciate the underlying structure. Understanding can come from systems thinking, in particular appreciating systems engineering, systems synthesis or systematic problem solving. It is also the author's observation that most people who use risk management in the workplace just go through the motions, cookbook style, according to the steps listed at the start of this chapter. This is aided and abetted by extensive use of checklists, standard forms, software packages, procedures, etc., which is encouraged in some of the risk management literature; filling in forms or completing checklists becomes risk management. People get numbers and take actions and make decisions, but do not understand the underlying problem-solving nuances of what they are doing. This lack of understanding has been observed by the author to lead to risk management results being unwittingly unusable for their intended purpose; if people do not understand what they are doing, then wrong risk management results could be anticipated to follow. However, it is the author's belief that, if risk management is looked at from the structure offered by systematic problem solving, such people would not only get better actions and decisions, but would also understand what they are doing. This view supports an argument for teaching systems engineering or systematic problem solving to all students.

Many believe that the insight gained through the discipline of going through the risk management steps can be more useful than any numbers that result. Accordingly, it is important to get the content of the steps correct, something which is awry in much of risk management as practised and in the risk literature.

4.3 PART A OUTLINE

The following chapters in Part A go through each of the risk management (RM) steps.

The first risk management step (*Definition and Context*) establishes the basis for the rest of the steps. Part of this is establishing the value system by which risks will be evaluated, the objective function(s) and constraints influencing the selection of the response, and a model for converting inputs to outputs.

Risk Source Identification requires the same type of thinking as generating ideas/alternatives in problem solving. A certain amount of creative thinking is required in this step for all but the most straightforward situations.

The analysis (*Analysis and Evaluation* step) may be carried out qualitatively or quantitatively using whatever technique and approach is appropriate.

The evaluation (*Analysis and Evaluation* step) is carried out with respect to the previously stated value system. This might involve some ranking and prioritizing of outputs, and even the dismissal of outputs regarded of little effect according to the value system.

The response (*Response* step) involves working with the inputs and input-output transformations according to the objective function(s) and constraints established in the *Definition and Context* step.

Chapter 5

RM step – definition and context

5.1 OUTLINE

Similar to systematic problem solving, part of definition and context thinking establishes the situation and value system of the stakeholder for whom the risk study is being undertaken. The value system and situation determine:

i. The measurement scale particulars – related to {output magnitude; output likelihood} pairs – for risk, for example, {low, medium, high}. Magnitude might be defined relative to some desired or base level, rather than an absolute magnitude.
ii. The objective function(s) by which the feedback adjustments are selected in the *Response* step – for example, lowest risk, minimum cost.
iii. The constraints restricting possible adjustments in the *Response* step – for example, some adjustments may be unacceptable.
iv. A model which provides the input-output transformation needed in analysis.

Different attitudes to risk (including what are popularly called 'aversion', 'neutral' and 'seeking' or similar terms) will lead to different measurement scale particulars for risk. Ideally, over time, an industry movement to agreed quantitative scales, rather than qualitative scales, might be attempted in order to allow better communication.

As synthesis, an objective function(s), model and constraints are naturally present (Carmichael, 2013a). The objective function(s) provides the basis for the adjustment choices made in the *Response* step. However, the need and use of an objective function in most risk publications does not appear to be appreciated because of the lack of realization that risk management is synthesis. Typically, risk is assumed in a downside sense, and hence minimization is usual if risk is part of the objective function. However, a more general view is that risk can have a downside sense or an upside sense. Objective function(s) and constraints are selected from the point of view of the stakeholder, for whom the risk study is being undertaken. Different people/organizations will select different objective function(s) and constraints,

DOI: 10.1201/9781003343592-6

and this will lead to differing adjustments in the *Response* step. The model provides the input-output transformation needed in analysis.

The underlying measure of risk has a particular scale based on the value system of the stakeholder for whom the risk study is being undertaken, together with the situation. The presence of the value system introduces subjectivity into risk management. Along with the situation, it also means that risk management is particular to each stakeholder at a point in time and situation, and therefore general statements on risk related to any matter cannot be made unless the value system and situation used are common to many people/organizations; for example, it cannot be said in general terms that one project delivery method or conditions of contract is '*riskier*' than another, even though such claims are prevalent in the literature (Carmichael, 2000 and Chapters 10 and 11). Different people/organizations characterize (including, in some cases, ranking and prioritizing) risk differently.

5.2 SUMMARY

In summary, in this definitional step, the following are established:

- The context of the risk study:
  - From whose point of view is the risk study being undertaken – client, consultant, contractor, public, …? This is referred to in this book as the stakeholder.
  - What is the nature of the risk being studied - safety, cost, production/schedule, …? This will prompt whatever background studies and data are needed.
- The way risks are to be evaluated in the *Analysis and Evaluation* step. (For example, outputs that impact humans might be regarded as primary risks, while outputs related to budget and operations might be regarded as secondary risks.) This might include acknowledgement of public opinion, perceptions, legal requirements and stakeholder background.
- The objective function(s) by which the adjustments are to be chosen in the *Response* step. (For example, minimum initial and ongoing cost.) The objective function(s) can include something relating to an end state (time or space).
- Constraints in the *Response* step. For example, certain adjustments may be unacceptable.
- A model which provides the input-output transformation needed in analysis.

Of the above bullet point items, those which reflect the value system of the stakeholder introduce subjectivity into risk management.

Note that it is an objective function(s) which is required (as in any optimal synthesis) for the *Response* step. This should not be confused with the popular management or lay over-usage of the terms 'objectives', 'goals' or 'aims', which are broad terms not really meaning anything useful, but sounding as though they are important. Objective function(s) result from the value system of the stakeholder. These may be difficult to establish for organizations, as opposed to individuals.

Carmichael (2020b,d) comments that the dilemma with using the terms 'objective', 'goal' or 'aim', is that these terms are used very loosely by most people, and extremely loosely in the managerial literature. For example, consider a project, which comprises designing and constructing a building; here the project's 'objectives' are commonly heard to be stated in terms of the project's end-product (typically desired qualities of the building) and not the project, or alternatively a project's 'objectives' and its scope (the designing and constructing work) are spoken of interchangeably (Carmichael, 2004). (Neither loose usage assists project management; rather the usages demonstrate the general lack of understanding of what management is, and the role objective functions play in management and decision making generally. This loose usage of terminology also highlights people's confusion over the difference between planning and design, even though both belong to synthesis, along with management.) The term 'objective' in lay and dictionary usage has imprecise and loose meanings, and almost certainly is different to that needed for proper decision making. The terms 'goal' and 'aim' also suffer. 'Goals' and 'aims' usually refer to end states (at an end time, or at an equivalent extremity of a spatial variable) and say nothing about what happens between time now and the end time (or equivalent spatial values). This inadequacy is noticeably highlighted if optimum control systems theory (Carmichael, 2013b) is studied; here the objective functions have two parts – something relating to an end state and something related to what happens up to the end time – both the trajectory and end state are important (Carmichael, 2004, 2013a).

Chapter 6

RM step – risk source identification

6.1 INTRODUCTION

Note: The term identification is used in the risk management literature to mean unearthing or uncovering, and not in the systems modelling sense.

Identifying inputs (risk sources) requires the same type of thinking as generating ideas/alternatives in problem solving. Some creative thinking is required in this step for all but the most straightforward situations; idea-generating techniques common to problem solving can be used here.

Some of the confusion over the meaning of risk can be seen in how this step is treated loosely in the literature – risk source and risk are commonly addressed as if they are interchangeable (Chapters 3 and 4). Strictly, this risk management step involves identifying inputs (risk sources), yet the step gets labelled in most publications as risk identification (equivalently, identifying something about the output), and many people believe that they are actually identifying something related to outputs (namely, risks), or do not appreciate the difference. The question that follows is: Why is it that people can quite happily use the terms risk and risk source interchangeably, when they clearly mean different things? Part of this may be laziness, but mostly, the author believes, it is due to people not understanding the difference, much like the lay misunderstanding of the difference between cement and concrete. Commonly, risk publications do an inconsistent treatment of identification. The confusion between risk source and risk gets perpetuated in the literature and in practice, and nobody seems to question the wrongness of interchangeably using the terms risk source and risk. To understand this identification step, dictionary definitions of risk need to be thrown away.

In health and safety, for example, risk relates to injury, illness or death, while risk source relates to whatever initiates this, for example, a hazard or an incident. Risk is not the initiator (even though loose lay usage may have people think it is). In safety matters, when risk management is done properly, this identification step might be referred to as hazard identification.

Risk sources are upstream inputs, while risks relate to downstream outputs (Figure 2.1a). A risk is not a risk source; an output is not an input (except for closed-loop systems). A risk arises as a result of a risk source.

People confuse the two. Possible outputs may be envisaged in the identification step, but only as a means to finding inputs that may produce such outputs (cause and effect). At this risk management step, the output characteristics are unknown, and the stakeholder is attempting to identify inputs.

There may be a combination of multiple sources, linked multiple sources, chain of sources or interrelated sources that lead to a downstream output, such as a failure or collapse. These sources may also be independent. A single source may not be sufficient by itself to lead to a downstream output. An example combination is one involving inadequate design, hardware failure and human error.

When there are competing interests and influences from others, for example, in contractual dealings or in sports and games, this introduces complications into risk source identification. In such situations, each's thinking depends on the other's thinking.

In an ongoing project, risk management and hence risk source identification is carried out on an ongoing basis, and not just once at the start of a project.

Effective identification relies on a working or intimate knowledge of the subject area, some imagination and some logical or systematic thinking. Risk sources are developed for the particular risk study situation at hand. The identification of risk sources can be facilitated with the help of:

- Checklists (experience- or knowledge-based; structured or unstructured).
- Interviews and questionnaires (including market research).
- Group idea generation techniques such as brainstorming.
- Experience (personal, corporate and industry), both good and bad. Repeating situations present the possibility of more readily identifying sources.
- Analysis methods (scenario, what-if, fault trees, event trees, computer-based,...).
- Past records.
- Flow charts.
- Specialist expertise.
- Systematic problem solving.

However, this list is not exhaustive. Other, more specialized approaches might be used. It is important to compile as many relevant risk sources as reasonably possible. Then Pareto-type thinking might be invoked, in that it may only be necessary to take forward the more important 20% of risk sources, with the other 80% considered to contribute little to the results of a risk study. This may mean, for example, that only a handful of risk sources are taken further for more detailed consideration.

In developing a list of potential sources, using a group of people from multidisciplinary backgrounds helps eliminate the tendency most people have for blinkered vision. Groups are able to make contributions from the many viewpoints represented. Individuals tend to think along certain paths based on their backgrounds and may overlook matters that other people might see. All people have 'blind spots'. People see what they want or expect to see, and hear what they want or expect to hear.

For example, in considering a safety issue in a manufacturing plant, a group might be set up composed of:

- People with equipment design and production skills.
- People familiar with the 'work face' operation, possibly people who work with the equipment on a regular basis.
- A facilitator with the ability to extract the knowledge, experience and creativity from the other group members.

Group membership is chosen to reflect the particular situation.

Counter to the benefits of groups, 'groupthink' is a process sometimes suffered by groups wherein there is pressure on individual members to conform. Groupthink is to be avoided and the group is to be encouraged to consider a wide range of possibilities.

In developing a list of risk sources, categories of risk might be used. A category describes the type of risk, for example, a risk related to health issues is put in the category labelled 'health risk', while a risk related to finances is put in the category labelled 'financial risk'. This is for compartmentalization convenience only, and does not signal a different treatment of risk. However, some of the literature on risk management regards (wrongly) these different categories as presenting risk differently.

Checklists are popularly used, but tend to be generic and need supplementing for particular situations. Each subsequent risk study adds something to the checklist such that over time the checklist becomes more and more comprehensive, with lesser chance of some source being missed. Sources may be listed under generic headings or categories such as: commercial; legal; economic; human; natural/environment, health; political; technological; and managerial. Generic categories of outputs might be considered in working backwards to initiating sources.

Inputs that lead to disasters may go unidentified because of their low probability of occurrence, while commonly present inputs, particularly those that directly affect people, are continually to mind. However, the lack of focus on low likelihood sources (and which might lead to high severity risks) may undo the risk management attempts. Risk sources that follow past habits can be more readily identified than something unusual or something that happens irregularly or infrequently. For example, gross movements in currency values resulting from an unusual economic condition may not be foreseen and hence not identified, while typical fluctuations in currency values would be identified. Similarly, a one-off disaster may not be anticipated, yet historical weather extremes would be. And without identifying a risk source, no management can ensue.

For uncluttered thinking, it is recommended that the initial focus is on identifying the nature of the risk sources, following which their magnitudes and likelihoods can be given attention. While identifying risk sources, a need for supplementary data or studies may arise.

The magnitudes of inputs might be established through judgement, past records, experience, industry practice, expertise, education, literature, interviews, idea-generation techniques, market research, experiments and/or historical or collected data. Likelihoods or frequencies of occurrence of inputs may come from observed frequencies, deductions from mathematical models, and/or measures of a person's subjective degree of belief of relative likelihoods upon which the practician is prepared to base a decision, as outlined more generally in Benjamin and Cornell (1970) and discussed below. With the last method, different people will come to different likelihood assignments because of their differing experiences and reasoning, unless a large amount of observed data are available to all people.

The reliance that can be placed on probabilities will depend on the sample size (data availability and documented events) and the basis of the quantification, which for small data sets may have to be logical reasoning or expert judgement. For example, weather events may be the subject of extensive records, while the uniqueness of many projects may work against the availability of much data.

Scale types discussed below are outlined in Chapter 2.

6.2 QUALITATIVE APPROACH

If a qualitative approach is adopted for inputs, then words and descriptive explanations are used involving an ambiguous magnitude scale such as {insignificant, minor, moderate, major, catastrophic} and an ambiguous likelihood scale such as {improbable, remote, occasional, possible, frequent} or {almost certain; likely; moderate; unlikely; rare}. Usually, a sort of translation is given of what these terms mean alongside the terms, but since English is used and lay English is imprecise, their meanings may still not add clarity – a case of connotations versus denotations. This represents an unresolved dilemma with using a qualitative approach. There appears to be no standardization of measurement scale particulars for magnitude and likelihood for both inputs and outputs. People also seem to guess, rather than reason, where their particular case lies on the scales. Ideally, the magnitude scale should approximately reflect real magnitudes. Also, ideally, the likelihood scale should approximately reflect probabilities if such probabilities can be established. However, in many applications, this does not appear to be the case. If the probability of a source cannot be established, then a verbal scale based on opinion may make the whole risk study pointless. Ordinal scale types (Chapter 2), giving order or ranking, but where the absolute size has no meaning, are commonly used for both magnitude and likelihood. Such scale types permit no more than medians and percentiles to be calculated. Ideally, over time, an industry movement to a quantitative approach might be attempted in order to allow better communication.

Some people feel more comfortable dealing with numbers as opposed to a purely descriptive/qualitative approach, while some people do not. However, the allocation of percentages (numbers), for example, can give people an unjustified faith in the risk study.

In some safety areas, likelihood scales may take the form of expressions such as {has happened at our facility yearly; has happened at our facility in its operational life; has not happened at our facility but has happened in our company or industry sector; has not been heard of in our industry}. These descriptors tend to have more meaning for people who work at the facilities and they can reduce the level of subjectivity in the assessment.

It has also been observed that although these scales refer to inputs, people may unknowingly be interpreting them in terms of any outputs, yet there may not be a one-to-one relationship between inputs and outputs. This produces a further dilemma when one input can have more than one output, for example, safety-related, environment-related and finance-related, and all these will have different scales.

Many writers use the terms 'probable' and 'improbable' to describe a 'reasonable likelihood of occurrence' and a 'low-level likelihood of occurrence', respectively, or similar. However, the choice of the terms 'probable' and 'improbable' (with lay meanings) is not desirable when describing likelihoods and probabilities (which have a well-defined technical meaning), and alternative terminology should be used. Similarly, the term 'possible' is used imprecisely and in lay terms, and often its meaning cannot be distinguished from the use of the term 'probable'. The term 'probable' should only be used in its correct technical sense and not a lay sense. Similarly, 'possibility' should be reserved for references about a variable's sample space and not be used in a lay sense.

6.3 QUANTITATIVE APPROACH

If a quantitative approach is adopted, numbers are used along with a magnitude scale, for example, in dollars or durations, with ranges from small to large, and a likelihood scale of probabilities, with ranges from small to large. Ratio scale types (Chapter 2), implying meaningfulness to the absolute values and the intervals between values, and where a value 0 implies the absence of the construct, are used. All statistical calculations may be carried out with a ratio scale type.

The sources of probability include (Tribus, 1969; Benjamin and Cornell, 1970; Ang and Tang, 1975):

- Observed frequencies.
- Deductions from mathematical models.
- Measures of a person's subjective degree of belief regarding possible matters.

The first bullet point, namely observed frequencies, requires a data set from which relative likelihoods can be calculated.

The last bullet point, namely a subjective degree of belief, refers to a person's judgement about something unknown, for example, the probability of bedrock at a certain depth, or the likelihood of completing a project in a given number of days. It reflects the person's belief in the relative likelihoods that apply. Another person, with a different background, may think differently about the relative likelihoods that apply.

A subjective degree of belief might be expressed as a probability (for example, the probability that the volume of earth is greater than some amount is 0.5), a percentage chance (for example, there is a 50–50 or 50% chance that the road will be congested), or a believed probability distribution (for example, an Erlang distribution with a stated shape factor is a good representation of work activity times – Carmichael (1987, 1989b), Carmichael et al. (2014), Carmichael and Mustaffa (2018), Carmichael et al. (2019b).

With enough beliefs expressed, a (subjective) cumulative distribution function can be plotted by joining the dots.

Subjective likelihoods are the basis by which utility curves or functions are constructed. Utility curves or functions are obtained by posing lotteries, in terms of a preference within a decision tree, to a person. The lottery is adjusted in response to the previous belief expressed by the person. At some point, it can be established where a person is indifferent between two lotteries.

Generally, people are uncomfortable with probability because of their lack of educational background in the subject. By comparison, a qualitative approach is easier to explain to a wider audience.

Qualitative inputs lead to qualitative outputs, while quantitative inputs lead to quantitative outputs.

Blends of qualitative and quantitative approaches are also used. Interval scale types, where the differences between measures, rather than the absolute values of the measures, is important, are used. With interval scale types, only a few statistical tools cannot be used.

6.4 ACHILLES HEEL

If an input is not identified, then risk management related to that input cannot proceed. This is *the Achilles heel of risk management*. The one input which has failed to be identified may bring about downfall. Aesop's Fable of the one-eyed doe provides: *Trouble comes from the direction we least expect.* Some well-documented historical building failures also example this. It is difficult to ensure that all possible risk sources have been identified. Current practices in input identification tend to be ad hoc, rather than ordered or structured. Ideally, something along the lines of a hierarchical breakdown from system to subsystem to sub-subsystem and so on, event tree, fault tree, cause-and-effect diagram, work study, or related structured approach would be better to use, because this should produce a more comprehensive list of risk sources compared with any ad hoc approach. However, this does not appear to be done in practice, possibly because of the restricted (non-systems) backgrounds of the persons doing risk studies, and because many people are not systematic in the way they think about things.

Learning from the past may be difficult – the documentation of failures is a sensitive issue because most people prefer to publish only achievements, as a result of legal liabilities and the reflection of failures on a person's/organization's reputation. Voluntary documentation of failures is thus rare.

The question that arises then is: What is the value of risk management if something can occur that has not been foreseen? 99% of all risk sources might be identified, but what of the 1% that is missed? Even the risk management steps themselves are a potential source or error, and therefore can lead to exposure. This raises an interesting issue. Should risk management of the risk management steps be undertaken (and continuing circularity)?

6.5 TOOLS ASSISTING IDENTIFICATION

The following tools are popularly used to assist identification.

6.5.1 Experience

'One learns from one's own mistakes' is a common saying. People tend to learn more from their own mistakes rather than from others' mistakes, because perhaps they feel the impact of such mistakes more and the mistakes leave more lasting impressions. Whether people are also able to analyse such mistakes in a better way, because of the personal involvement, is hard to say, even though they may be more familiar with the circumstances. In occupational health and safety matters it may be mandatory to document every unsafe event in sufficient detail for the benefit of others. Documentation of failures is a sensitive issue.

Expert systems and artificial intelligence approaches can assist in encapsulating experience.

6.5.2 Event trees, fault trees

Event trees allow the logic to develop in going from an initiating event (input) to outputs (consequences) (Figures 6.1 and 6.2). In some cases, thinking may start with possible outputs and then work backwards to the source (Figure 6.3).

Chapter 20 deals with event trees and fault trees in greater detail.

Other methods include events and conditions charting, hazard and operability studies (HAZOP), and failure mode and effect analysis (FMEA).

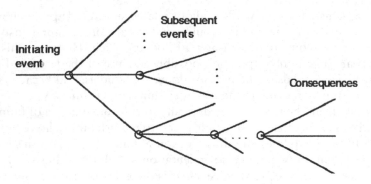

Figure 6.1 Input (initiating event) leading to outputs (consequences, outcomes) – event tree.

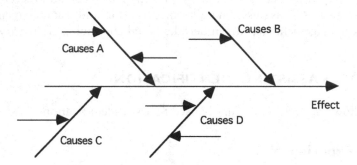

Figure 6.2 Cause-and-effect, Ishikawa, fishbone or affinity diagram; cause – inputs (contributing events); effect – output (consequence, outcome).

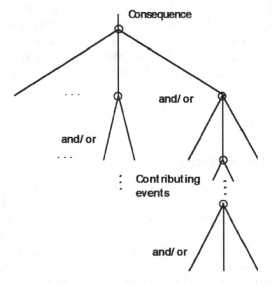

Figure 6.3 Output (consequence, outcome) – caused by contributing inputs (events) – fault tree.

6.5.3 Checklists

Checklists provide a useful memory prompt (an aide memoire). They may be based on past work, or databases developed by specialist groups or specialist industries. Some example checklists follow. The items in these checklists could be anticipated to be broken down into finer detail to deal with specifics.

The breakdown structure of Table 6.1 is an example checklist of risk sources that can adversely impact an infrastructure project. All sources may not be relevant for all projects. It is a matter of determining applicability in each circumstance.

Table 6.2 examples another possible way of categorizing project risks.

Table 6.1 Example project risk sources

Externals	Contracts/ Procurement	Resources	Internals
• Market, demand, changes. • Acquisitions, delays. • Community expectations. • Political input. • Heritage. • Weather and effects. • Public utilities, delays, costs. • Environmental matters, constraints. • Bureaucratic delays. • New legislation, regulations (statutory changes). • Technological change. • Wars, law and order. • Geographic isolation.	• Contractor ability. • Workmanship. • Contractor default. • Dispute resolution mechanisms. • Latent conditions. • Interfacing of contracts. • Industrial disputes. • Claims.	Money • Acquisition costs. • Inflation. • Availability. • Funding, delays. • Interest rates. • Commodity prices. • Foreign exchange. People • Skill level. • Availability. Materials • Suitability. • Supply. • Availability. • Geotechnical. • Plant/equipment • Reliability, breakdown. • Availability. Escalation – rise and fall in prices.	Objectives • Inadequate. Time Surveys Estimating • Accuracy. Specification • Inadequacy. Design • Adequacy, failures. • Defects. Construction • Failure of temporary structures. • Workmanship. • Techniques. • Defects. Policies • Change of (quality) standards. • Unclear objectives. Management • Indecision.

Table 6.2 Examples of project risk sources (based on a New South Wales Government publication)

Planning and feasibility stages	Project delivery stages
Commercial and strategic	**Procurement and contractual**
• Competition.	• Contract selection.
• Market demand levels.	• Client commitment.
• Growth rates.	• Consultant/contractor performance.
• Technological.	• Tendering.
• Stakeholder perceptions.	• Negligence of parties.
• Market share.	• Delays – weather, industrial disputes.
• Private sector involvement.	• Damages and claims.
• New products and services.	• Errors in documentation.
• Site acquisition.	• Force majeure events.
	• Insurance and indemnities.
Economic	
• Discount rate.	**Construction and maintenance**
• Economic growth.	• Buildability.
• Energy prices.	• Contractor capability.
• Exchange rate variation.	• Design and documentation.
• Inflation.	• Geotechnical conditions.
• Demand trends.	• Latent conditions.
• Population growth.	• Quality controls.
• Commodity prices.	• Equipment availability and breakdowns.
	• Obsolescence.
Contractual	• Industrial action.
• Client issues.	• Materials availability.
• Contractor issues.	• Shut-down and start-up.
• Delays.	• Recurrent liabilities
• Force majeure events.	• Health and safety.
• Insurance and indemnities.	• Accident, injury.
• Joint venture relations.	• OH&S procedures.
	• Contamination.
Financial	• Noise dust and waste.
• Debt:equity ratios.	• Disease.
• Funding sources.	• Irradiation.
• Financing costs.	• Emissions.
• Taxation impacts.	
• Interest rates.	**Human factors**
• Investment terms.	• Estimation error.
• Ownership.	• Operator error.
• Residual risks for Government.	• Sabotage.
• Underwriting.	• Vandalism.
Environmental	**Natural events**
• Amenity values.	• Landslip/subsidence.
• Approval processes.	• Earthquake
• Community consultation.	• Fire.
• Site availability/zoning.	• Flood.
• Endangered species.	• Lightning.
• Conservation/heritage.	• Wind.
• Degradation or contamination.	• Weather.
• Visual intrusion.	

(Continued)

Table 6.2 (Continued) Examples of project risk sources (based on a New South Wales Government publication)

Planning and feasibility stages	Project delivery stages
Political	**Organizational**
• Parliamentary support.	• Industrial relations.
• Community support.	• Resources shortage.
• Government endorsement.	• Scheduling.
• Policy change.	• Operational policies.
• Sovereign risk.	• Management capabilities.
• Taxation.	• Management structures.
	• Personnel skills.
Social	• Work practices.
• Community expectations.	**Systems**
• Pressure groups.	• Communications or network failure.
Project initiation	• Hardware failure.
	• Linkages between subsystems.
• Analysis and briefing.	• Software failure.
• Functional specifications.	• Policies and procedures.
• Performance objectives.	
• Innovation.	
• Evaluation program.	
• Stakeholder roles and responsibilities.	
Procurement planning	
• Industry capability.	
• Technology and obsolescence.	
• Private sector involvement.	
• Regulations and standards.	
• Utility and authority approvals.	
• Completion deadlines.	
• Cost estimation.	

Table 6.3 shows an example checklist for construction-type activities on a project.

Table 6.4 gives an example checklist for risk sources in procurement.

With checklists, it is always possible that a source may have been overlooked. Ideally, the checklist should be developed independently of any analysis and evaluation, as well as independently of magnitude and likelihood estimates. The same applies to techniques such as brainstorming which should strictly be idea generation, with no analysis and evaluation until after the brainstorming session is complete.

In terms of trying to capture as many sources as possible in the checklist, some recommended practices are:

- The approach to developing the checklist is done systematically, much like a work breakdown structure (WBS) is developed in planning (Carmichael, 2006), that is, system to subsystem to sub-subsystem and so on.

Table 6.3 Example checklist for construction activities

Construction elements	Variations	Weather	Equipment delay/ failure	Crane delay/ failure	Pump delay/ failure	Material delay	Contractor failure	Labour issue	Resource delay	etc.
Sheet piling		x	d/f	d/f		x	x		x	
Excavation		x	d/f					x		
Foundations		x	f	f	d/f	x	x	x	x	
Corewall	x	x				x		x	x	
Machine room		x		f	f	x		x	x	
Floor slabs		x		f	f	x		x	x	
Roof				f	f	x		x	x	

- People with as many different backgrounds as possible are engaged, much like in idea generation in problem solving.
- Checklists are updated with each new experience.
- Information obtained from other identification tools is incorporated. These tools include: input from experienced personnel; published information or past projects (including variations and lessons learnt); brainstorming; peer review; what-if; and trialling the checklist on previous (completed) projects.

There is a view that the value of a checklist is reduced by including sources which later turn out to be not important. Some judgement and experiential input may therefore be necessary for producing a checklist.

As a cautionary note, if followed without question, checklists may hinder imaginative thinking which is necessary for identifying the unusual risk source. Checklists may constrain 'thinking outside the square'. Checklists can encourage a lazy attitude to risk management. Some people see them as another tick-the-box chore, or assume that if something is not on the checklist (written by someone else) then it is not an issue, and so it can be ignored.

6.5.4 What-If analysis

Asking the question 'what-if?' is a generally applicable approach that overlaps or could be used in conjunction with other approaches to identification. Some imagination or prompt is needed to answer the question.

6.5.5 Deviations, sensitivity

A related approach is to establish what is usual or normal performance, and deviations from this are then considered.

Table 6.4 Typical risk sources in procurement (based on an Australian Government publication)

Buyer-related	Seller-related	Contractual relationship-related	External-related
Availability and cost of finance	Financial security	Optimum pricing levels	Public perceptions of costs and benefits
Source of funding	Cost-estimating procedures	Payment schedules	Currency fluctuations
Project management competence	Misinterpretation of requirement	Cost control mechanisms including price variation provisions	Orphan technology
Availability or resources	Project management competence	Contractual liabilities	Security
Impact on other projects/ programs	Manufacturing capability	Adequacy of insurance	Non-compliance with contractual conditions
Effectiveness of program management	Availability of resources	Quality requirements	Legislative changes
Training	Shipping and delivery	Ergonomic considerations	Environmental changes
User resistance to new technology	Installation, testing and commissioning	Ownership of intellectual property	Political considerations
Inadequate contractual arrangements	Inadequate offer		
Selection of wrong supplier	Integration of wrong supplier		
Use of inappropriate procurement method	Reliability, availability, maintainability		
Inadequate evaluation of offers	Fraud		
Inefficiency	Collusive tendering		
Fraud	Collusive tendering		
Inappropriate specifications	Theft or criminal action		
Conflicts of interest	Waste, error		

A sensitivity analysis looks at small changes or small deviations from the usual or normal. This is compared with a what-if analysis which has no limitations on the size of the deviations (Carmichael, 2013a).

6.5.6 A systematic framework

So as not to leave something out, it can be beneficial to systematically work through the situation at hand. For example, in considering what could go wrong in the construction of a building, the building construction is first broken down into its components – substructure construction and superstructure construction. These are then broken down further, for example, the substructure construction is broken down to excavation, dewatering, shoring, forming, steel fixing, pouring concrete, etc. These are then broken down further, and so on. That is, the overall task of constructing a building is broken down level by level to its elemental tasks. Then an assessment is made as to what could go wrong in performing each of the elemental tasks. It is akin to the systematic framework adopted in work breakdown structures (WBS) used in planning (Carmichael, 2006). A similar system-to-subsystem-to-subsubsystem-… breakdown can be adopted for other situations. For example, an examination of risks in marketing might look at the components of the marketing mix, namely product, distribution, price and promotion.

6.5.7 Process and flow charts

Outline process charts and flow process charts used in work study (Carmichael, 1997b, 2013a) can be useful in delineating component work practices. An assessment can then be made of potential risk sources for each component work practice.

Such charts are examples of what many people group as flow charts. The flow may be in terms of materials, mechanical components, people, money, etc. Also included in this group are the networks used in project planning and scheduling, traffic networks showing flows of vehicles, and communication networks.

6.5.8 Audits

An audit is a check against a set of standard criteria, regulations or procedures. It is an inspection, correction, verification and certification process. It is most often thought of in connection with business accounts but has wide applicability to all operations. Although not directly intended for risk management, it can be useful for uncovering information that feeds into the *Risk Source Identification* step.

Audits are a basic part of quality assurance procedures used in many organizations.

6.5.9 Historical perspective

Sources present in past situations may also be present in the current situation. Past data are commonly held, for example, on accidents, critical incidents and workers compensation. Maintenance and equipment damage records are also commonly held. Common failure modes in past computing systems and past software development are noted. Present-day quality systems tend to provide an oversupply of information.

For example, in a project involving the supply of hardware and software for building management, security and photo-ID for a prestige building, common risk sources are: staff/resources unavailable when required; specification creep; hardware not available/not delivered on time; and time constraints.

6.5.10 Ergonomics

Ergonomics is the study of the relationship between the work being undertaken, the worker and the work environment. The behaviour of people and their work functions may be broken down into activities. Each activity may be examined for possible things that can go wrong while this work is being undertaken, such as a failure, injury and decreased production. The questioning practice adopted in Work Study (Carmichael, 1997b, 2013a) can be used, that is:

What (is achieved)?	Why (is it necessary)?
Where (is it done)?	Why (there)?
When (is it done)?	Why (then)?
By whom (is it done)?	Why (that person)?
How (is it done)?	Why (that way)?

Answers to these questions can then lead on to a consideration of risk sources.

6.5.11 Industry publications

Industry publications might assist in that they may have already listed many well-known situations. For example, with respect to construction hazards, the Construction Work Code of Practice (Safe Work Australia, 2018) lists what it terms 'high risk' activities:

- Falling more than 2 metres.
- Work on a telecommunication tower.
- Demolition of a load-bearing structure.
- Disturbance of asbestos.
- Alterations involving temporary support.

- Work in a confined space.
- Work near a shaft or trench.
- The use of explosives.
- Work near pressurized gas, chemical/fuel lines, energized electricity.
- Work in a contaminated or flammable atmosphere.
- Tilt-up precast concrete work.
- Work adjacent to a road, railway line, ...
- Work near powered mobile plant.
- Work in artificial temperature extremes.
- Diving work.

6.5.12 Brainstorming

Brainstorming is a technique used to generate ideas. It is based on encouraging a free flow of ideas, unconstrained by inhibition, prejudgement, criticism or evaluation. Emphasis is on generating as many ideas as possible, even ideas that, on later analysis, may turn out to be not important. All ideas generated have the potential to stimulate further ideas.

Brainstorming is a group technique where ideas put forward by one group member stimulate other group members to thought. The approach is particularly useful for new, unusual or innovative work, and for the initial development of checklists. In terms of identification, the phrase 'think the unthinkable' is relevant.

Basic rules or principles applying to brainstorming include: no judgement or criticism including not using 'blocking' phrases (anything preventing free thinking); all to participate; and quantity rather than quality.

The term 'workshopping' might be used by some people to describe processes involving brainstorming-type approaches.

Some people suggest that unstructured brainstorming can have its effectiveness improved by giving some logical or systematic framework to the exercise.

Some variants on brainstorming include the following:

- Where there is the potential for group members influencing, in a negative sense, or inhibiting other group members, a variant on the brainstorming technique isolates subgroups who then develop their ideas separately. All ideas are later shared. Subsequent isolated subgroup work and sharing may follow. This, or similar, may be referred to as the Delphi technique.
- Where the group leader is the only person who knows the actual situation, group discussion centres on the situation generally without group members knowing the exact situation. It is up to the group leader to focus the discussion.
- Discussion as to preferred ideas may be introduced and this alters the non-evaluative form of pure brainstorming.

Chapter 7

RM step – analysis and evaluation

7.1 INTRODUCTION

7.1.1 Analysis

Analysis takes the inputs (magnitude and likelihood) and converts these into outputs (magnitude and likelihood), as in systems analysis generally. The analysis may be carried out qualitatively or quantitatively. A qualitative analysis uses verbal models, in contrast to a quantitative analysis which uses mathematical models.

The scale types discussed below are outlined in Chapter 2.

If a qualitative analysis, an output magnitude scale such as {disruption, damage, loss of property, injury, death}, {negligible, minor, serious, severe, catastrophic} or {insignificant; minor; moderate, major, catastrophic} or {minor; critical; major; catastrophic}, and an output likelihood scale such as {rare, unlikely, moderate, likely, almost certain} or {frequent; possible; occasional; remote} might be used, along with descriptive explanations. Such scales can always be criticized for their ambiguous meanings.

Examples of qualitative magnitude scales for different situations could be in terms of the following: earthquakes – {no damage; people evacuate, small damage; people injured, part building damaged; heavy injury, all building damaged; death, building destroyed}; flooding/structures – {water damage; minor electrical/plumbing damage; major electrical/plumbing damage; minor structural damage, mechanical/electrical failure; major structural damage}; earthquake/human perception – {small vibrations; large vibrations; objects falling down; structure cracks; structure collapse}; and general – {disruption; damage; loss of property; injury; death}. All can be criticized for their ambiguous meanings.

For the qualitative case, the output magnitude scale will approximately reflect real values, while the output likelihood scale will approximately reflect probabilities.

If a quantitative analysis, output magnitude scales might be, for example, in dollars, durations, number and type of human injuries or number of

DOI: 10.1201/9781003343592-8

fatalities, money lost, delay extent or damage extent, while an output likelihood scale of probabilities is used.

Blends of qualitative and quantitative approaches are also used.

As with the scales for inputs, qualitative, quantitative and blend approaches use ordinal, ratio and interval scale types, respectively.

A quantitative analysis would generally employ some numerical approach, perhaps computer-based. It revolves around establishing quantitative estimates for outputs such as costs, times, … with associated uncertainty. See for example, Carmichael (1988, 1990), Balatbat et al. (2012), Carmichael and Balatbat (2011a,b), Carmichael and Edmondson (2015), Carmichael et al., 2016) and Jukic and Carmichael (2016). Multiple interrelated inputs require more care in how they are incorporated into the analysis.

It is unclear how people establish the scales for magnitude and likelihood, and whether it would be more preferable to have an approximately linear scale, a nonlinear scale or something else. Looking ahead to that involving evaluation and taking a qualitative approach, the popular use of matrices drives a 'step-function' or discrete view of what is happening (rather than a continuous linear or nonlinear one). (Matrices are discussed below.) Some people try to iron out the steps by having larger and larger evaluation matrices (for example, up to 10×10). However, this can add to the confusion, because people appear to be able to only delineate qualitatively into 3 to 5 divisions. Asking someone to refine into 10 different divisions can give a false sense of being more accurate. In practice, up to about 5 divisions appears common. Industry seems to have shaken down to this.

With a qualitative approach, and being based on verbal models, there is always the issue of people's different perceptions of the words, and connotations versus denotations. There appears to be no standardization of scales for magnitude and likelihood for both inputs and outputs. People also seem to guess, rather than reason, where their particular case lies on any scale.

7.1.2 Evaluation

The {output magnitude; output likelihood} pairs are located on the particular measurement scale for risk appropriate to the stakeholder for whom the risk study is being done. Each pair could be anticipated to be located or mapped differently for different stakeholders because of their different value systems and situations. The use of tables, matrices or plots with 'axes' of output magnitude and output likelihood, and entries of risk levels, might be used to simplify this, as discussed below. The fitting to a particular measurement scale for risk is commonly called evaluation (or assessment) in the literature, while the literature may overlap the *Analysis and Evaluation* step and the *Response* step. An example qualitative scale for risk is {low, medium, high} or some severity descriptors.

7.1.3 Of interest

Some interesting approaches have been personally observed in analysis/ evaluation carried out by practitioners:

(i) For the probability or likelihood to carry through from input to output there has to be a direct relationship between the two. However, a common assumption made by people is that the probability of an input (for example, rainfall) is the same as the probability of the output (for example, flooding) without investigating whether this is true or not.

(ii) Because of the imprecision involved in qualitative analyses, people adjust or manipulate magnitudes and/or likelihoods in order to achieve wanted results.

(iii) Risks are sometimes established, usually without justification, as the product of output magnitude and output likelihood. This is akin to a 'risk neutral' attitude, but curiously most people/organizations are 'risk averse' and not 'risk neutral', and so taking the product may not be valid. This practice possibly comes about through representing risk within a matrix with (numbered) 'axes' of output magnitude and output likelihood (for example, each 'axis' division being numbered from 1 to 5 as exampled below), having this matrix symmetrical or nearly symmetrical about a diagonal, and doing risk management cookbook or do-it-by-numbers style. However, the practice ignores the value system and situation espoused in the *Definition and Context* step, ignores the properties of the measurement scale particulars for risk, output magnitude and output likelihood, and might be viewed as giving output magnitude and output likelihood equal weighting. It implies linear scales for output magnitude and output likelihood. It also can lead to anomalies such as the risk being the same for {low output magnitude; high output likelihood} and {high output magnitude; low output likelihood}, for example in the construction industry having the same risk associated with a crane topple as for slips and falls. It is remarked, however, that although the risks may be thought to be the same in this example, the response applied (in the ensuing *Response* step) might be different. (Asymmetry in the matrix can be introduced by having the 'axes' for output magnitude and output likelihood numbered differently. Or a formula for risk different to the direct product between output magnitude and output likelihood could be used.)

In terms of the relationship between the input probability and the output probability, consider some other examples. Example (a): The probability of a storm event is greater than the probability of the resulting overflow of drainage infrastructure. Other matters contribute such as surrounding ground conditions, infiltration, blockage within the infrastructure, and so on. Example (b): If a local currency devalues, an importer (buying in a foreign currency) will be impacted by the resulting increase in the cost of foreign goods. The probability is transferred directly. Example (c): The probability of Legionellae occurring in cooling water or the probability of Giardia or Cryptosporidium in potable water is higher than the probability of someone becoming infected.

Knowledge of the output that results from any analysis is useful, but some believe that the discipline of going through the analysis and evaluation and the insight gained of the nature and origin of risks is more useful.

7.2 ANALYSIS

Commonly, the term 'risk analysis' can be found in the literature, and reflects the general looseness of the literature on risk management. The expression is misleading, because what is being done in analysis (*Analysis and Evaluation* step) is converting inputs into outputs. No value system of the stakeholder has been invoked yet (in progressing through the risk management steps), and so risks have yet to be established, let alone 'analysed'. The evaluation follows on from the analysis (*Analysis and Evaluation* step).

Analysis might be undertaken in two stages. The first stage is a qualitative analysis that carries through to a subjective view of the output. The second stage is a quantitative analysis that presents a more definite view of the output. The qualitative analysis might be undertaken in order to obtain an overall understanding; it assists in excluding (or screening) less important issues from being a part of a more detailed study. If no more is done in terms of analysis, at least a qualitative analysis is considered essential. As a result of a qualitative analysis, there may be no need for any quantitative analysis. For cursory risk management, only a qualitative analysis may be carried out. For something involving significant investment or money, a quantitative analysis would be more common. But at what point a qualitative analysis becomes insufficient and a quantitative analysis becomes necessary cannot be said with any definiteness. A qualitative-quantitative blend of analysis is also possible.

There is a view that because probabilities cannot be established with any real accuracy, being often based on judgement and subjectivity rather than historical or collected data (Chapter 6), performing a quantitative analysis cannot be justified in most cases. There is also the issue of how well the model used for analysis represents reality and the availability and adequacy of data. In many cases, GIGO (garbage in, garbage out) is seen in published risk management practices. This could point to qualitative analyses being the more justifiable in many circumstances.

Circumstances in which a qualitative analysis might be employed include:

- When first examining a situation, for example, in the concept and development stages (phases) of a project.
- Where the present development or level of knowledge does not permit a complete quantitative analysis, including the situation where data are not available.

- Low budget and low resource situations where the expense of a more time-consuming quantitative analysis is not warranted.
- Where it is only needed to prioritize risks.
- Where qualitative information on risks is sufficient for decision-making purposes.
- Where it is anticipated that the level of risk will not justify a more detailed analysis.
- Where a quick view is sought, or time is limited.
- Where exact data are not available or required.
- Where the user is ignorant of quantitative techniques, through perhaps lack of education.
- For small issues.
- As a filter to a quantitative analysis.

In many situations, a qualitative analysis may be the only analysis carried out.

Descriptive likelihoods are used in a qualitative analysis compared with probabilities in a quantitative analysis. The talk and use of probability seems to create an insecure feeling in people, possibly because of their deficient backgrounds in mathematics; by contrast, using likelihoods does not seem to promote insecurity. Accordingly, a qualitative analysis gains ready acceptance among most people, as compared to a quantitative analysis.

A sensitivity study might be attempted through varying the input values slightly and observing the ensuing outputs, in order to establish confidence in any analysis. This can be done qualitatively, but more usually quantitatively. The reader is referred to Carmichael (2013a) for more detailed comment on sensitivity analysis.

A 'what-if' analysis asks the question, what if the input or something about the system is changed, then how will that affect the output. Refer Figure 2.1a. A sensitivity analysis does the same, but only small changes (± a few per cent) are considered. The analysis is done deterministically. Typically, only one input change at a time is investigated.

Sensitivity can then be ascertained as in Table 7.1. Output changes of magnitude bigger than the perturbing input changes or of an order of magnitude bigger than the existing output would commonly imply high sensitivity (to input or system changes). Small magnitude changes in the output would imply low sensitivity.

Table 7.1 Sensitivity conclusions

Change in input or system	Change in output	Sensitivity conclusion
Small	Small	Insensitive to change
Small	Large	Sensitive to change

Just as feedback occurs in systematic problem solving, so too does it occur in the risk management steps. Often identification is intertwined with analysis, evaluation and response and they all proceed hand-in-hand, sometimes deliberately but also commonly because practitioners do not understand the underlying system fundamentals of risk management. Nevertheless, each step performs a distinctive purpose.

7.3 EVALUATION

Evaluation might also be termed assessment in the literature. The same thinking in evaluation applies whether the preceding analysis has been done qualitatively or quantitatively.

The literature blurs the distinction between analysis and evaluation and often interchanges the two without realizing the error.

Evaluation takes the output pairs {output magnitude; output likelihood} and translates these to a particular measurement scale for risk, which is based on the stakeholder's value system. These might then be reported in some priority order, commonly from 'high' risk to 'low' risk.

To these descriptors of risk, an organization might attach some responsibility, for example: 'high' – senior management; 'moderate' – middle management; and 'low' – workface. But this is separate to the risk management steps treated in this book.

Evaluation only has meaning with respect to the stakeholder's value system. Too often it is seen that people go from analysis to evaluation and grab risk values out of the air, without reference to any value system. This reflects a lack of understanding of risk management. The commonly seen cookbook, do-it-by-numbers approach to risk management is a contributor to this.

For a qualitative approach, matrices, as mentioned earlier, are commonly used in evaluation. For example, the output magnitude 'axis' may be divided according to {insignificant; minor; moderate; major; catastrophic} while the output likelihood 'axis' may be divided according to {almost certain; likely; moderate; unlikely; rare}, giving a 5 × 5 matrix or 25 cells. Within each cell, the risk level corresponding to the stakeholder's value system is inserted and, for example, these might correspond to high, significant, moderate and low. The information inserted into the cells is the evaluation, for example as in Table 7.2. Both the matrix horizontal and vertical 'axes' and the cell information suffer from the imprecision of the English language and the case of connotations versus denotations. There is a lot of subjectivity in such an evaluation.

An alternative to using a matrix is to plot regions of risk levels separated by curves where there is a transition from one risk level to another, for the same two 'axes' – output magnitude and output likelihood, as exampled in Figure 7.1. The plot or risk chart contains the same information as a matrix with the same 'axes'.

Table 7.2 Example qualitative evaluation matrix. H – high risk; S – significant risk;
M – moderate risk; L – low risk

	Output magnitude				
Output likelihood	*(1) Insignificant*	*(2) Minor*	*(3) Moderate*	*(4) Major*	*(5) Catastrophic*
(1) Rare	L	L	L	M	S
(2) Unlikely	L	L	M	M	S
(3) Moderate	L	M	M	S	H
(4) Likely	M	M	S	H	H
(5) Almost certain	S	S	H	H	H

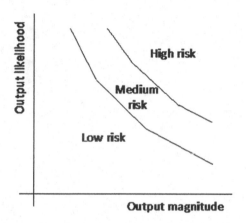

Figure 7.1 Example evaluation plot or chart.

There appears to be no standardization in industry of the form for a matrix or chart. Each is developed one-off for each different case for each different stakeholder.

Qualitative and part-quantitative approaches lead to risks in a relative sense. A quantitative approach uses absolute values for output magnitude and output likelihood, potentially giving risk values that are directly comparable.

Chapter 8

RM step – response

8.1 INTRODUCTION

The *Response* step might also be termed the control or treatment step in the literature. In health and safety matters, it might be referred to as applying controls. The use of 'control' here is the same as that in control systems theory, that is an influencing action, and not as popular management uses the term as 'containment'.

This is the decision-making step of risk management. Decisions (variously referred to as responses, control choices or adjustments) are made based on the objective function(s) and constraints delineated earlier in the *Definition and Context* step.

The *Response* step deals, where relevant, with influencing the inputs (magnitude and likelihood), or transformations from input to output (and hence indirectly influencing the magnitude and likelihood of outputs), through actions/decisions, by feedback, in order to address the risks just evaluated. Adjustments (the responses) to the inputs and input-output transformations are examined in terms of the earlier stated objective function(s) and constraints. Those adjustments which lead to minimizing (or maximizing, as the case may be) the objective function(s), while satisfying any constraints present (admissible values) are preferred. If the steps are done iteratively, the adjustments are examined similarly.

The generation of adjustments parallels that in systematic problem solving. Some creative thinking is required in this step for all but the most straightforward situations; idea-generating techniques common to problem solving can be used here.

Adjustments are only made if they favourably influence the value taken by the objective function(s) and do not lead to something inadmissible (in the sense of violating the constraints). There is also the 'do nothing' alternative, that is, nothing is changed, if applicable. Low or minimal risks (depending on the objective function(s)) may be accepted without further consideration.

For elementary situations, expected value or expected utility might be a sufficient objective function for establishing preferred responses. Expected value weights output magnitudes with their probabilities. Utility adjusts the value of something depending on whether the decision maker is a 'risk seeker' or 'risk averse' (Carmichael, 2013a, Carmichael et al., 2018).

The role of the stakeholder's value system leading to the objective function(s) enunciated in the *Definition and Context* step is not understood in the literature. It is not uncommon to see statements, made in isolation, such as select the response which: 'gives maximum risk reduction'; 'balances the cost of the response with the benefits flowing from the response'; 'gives the largest risk reduction at the lowest cost'; 'enhances opportunities and reduces threats'; is 'appropriate for the significance of the risk, cost-effective, realistic, agreed upon by all, and owned by one person'; or is based on 'the cost of rectifying any potential consequences versus the opportunities afforded by accepting the risk'. All such statements show a lack of understanding of risk management, but superficially sound impressive.

The literature also mixes objective functions and constraints, without appreciating their different roles. Constraints are never explicitly recognized as such in the literature. For example, a response may be termed 'not justifiable on economic grounds'. Occasionally, the term 'optimum' or 'optimization' is included to impress; however, without formally recognizing the components to the problem solving and optimization, this has no meaning. Optimization only has meaning with respect to an objective function(s), and this is never revealed.

Adjustments may occur at different points in time. Two timings of the adjustments may be recognized:

- *Immediate responding.* Adjustments are introduced in order to address the risk now.
- *Contingency responding.* Adjustments are developed to deal with any potential risks, but are only implemented should a prescribed input occur. Trade-offs may be considered in terms of making adjustments at different points in time and the different outputs associated with making adjustments at different times.

Adjustment choices might be categorized according to whether they involve, singly or in combinations:

- *Elimination* (including: removal, avoidance, transference*); the value of a (minimum) risk objective function goes to zero.
- *Reduction* (including: mitigation, part-transference*, sharing*); this leads to a lower value for a (minimum) risk objective function.
- *Leaving unchanged* (including: acceptance, retention, assuming); the value of a (minimum) risk objective function remains the same.

Related choices apply for upsides.

* As noted earlier, each stakeholder has a different value system; accordingly, for any given situation, the risk to one stakeholder is different to the risk to another stakeholder. That is, risk (as perceived by one stakeholder) is not preserved in any form of transference or sharing (with another stakeholder). Yet, it is frequently seen and heard that the total risk is regarded as a constant, capable of being divided into pieces, much like a cake being cut into slices, and various stakeholders being allocated slices, or stakeholders swapping slices. This error stems from loose thinking on what risk actually is.

Risk attitudes (seeking, neutral, averse) may influence the adjustment choices, but would already have been considered through the stakeholder's value system.

Common practice observed is to not recognize the role played by the objective function(s). In any synthesis, multiple adjustment choices are possible, leading to non-unique results; the non-uniqueness is removed through the presence of an optimality measure, namely the objective function(s) (Carmichael, 2013a). Generally, the literature is unaware that a complete formulation in optimization requires an objective function(s) against which the degree of optimality of an adjustment can be tested; rather, mention is made of responses feeling appropriate against indefinitely stated goals, commonly involving some indefinitely defined balance between costs and risks. No mention is ever made of an objective function(s), and hence how optimal any response is, or whether better (and admissible) responses might exist. The stakeholder is unaware of how good or how bad a response is. The elimination of as many risks as possible can be a mindset without regard to cost or any other measure.

Note: In this book, the term objective function(s) is deliberately used, and not the term objective. The use of the term objective in the popular management and risk literature is possibly in the sense of a goal or end-state and is used in a vague way to impress rather than help. Outputs are compared with the objectives in order to establish the difference between the two. The use of the term objective in this way is not to be confused with the use of the term objective function in this book (Carmichael, 2013a), it being the criterion by which the optimum values of the adjustments are established. The literature does not explicitly acknowledge the components of synthesis or explicitly acknowledge that synthesis is involved in risk management, even though qualitative expressions equivalent to systems objective functions and constraints may be stated as window dressing (but not used).

Organizations seem to be at ease with documents called risk registers or similar. Such registers are laid out in columns, with some but not all of the following information: input/source (description, magnitude, likelihood), output (magnitude, likelihood) (though typically the likelihood of the output is set equal to the likelihood of the input, but without justification), pre-control risk, control and post-control risk. Such registers can be the sole evidence of an organization's risk management practices. They demonstrate risk management by numbers or by cookbook and generally demonstrate that risk management is not truly understood. Safe Work Method Statements (SWMS) (Safe Work Australia, 2018) are like risk registers; however, here the complete removal of injuries and deaths is the purpose and hence can be much simpler in content that risk registers applied to non-health/safety outputs.

8.2 EXAMPLE ADJUSTMENTS

Some example adjustments are given here.

Example – contingencies. In planning, a contingency is another name for a cost reserve or a time reserve. It is intended to allow for KUKs (known unknowns). Contingencies commonly occur in cost estimates. Float in project programs is a form of reserve. A tolerance or 'space' in a performance specification is another example. Contingencies could be thought of as an example of reduction. Typically, contingencies are added as a percentage, for example, an extra 10% might be added to a project estimate. This percentage is based on whatever worked well last time, whatever is customary in that industry, or whatever seems right according to 'gut feel'. But the rationale of adding a percentage is not correct. Ideally, risk should be considered at the component level, and a contingency built up to the project level (Carmichael, 2006). But nobody seems to do this.

Example – document error. Courts commonly interpret contractual documents on a contra proferentum basis – that is, against the author of the documents, should there be errors, ambiguities or inconsistencies in the documents. That is, risk associated with errors in contract documents is generally with the owner/principal. One way of reducing some risk to the owner is to eliminate authorship of some of the contract documents. For example, by not including a bill of quantities, contractors may have to develop their own bill of quantities. Risk associated with errors in this document now rests with the contractor. Claims through errors in the bill of quantities no longer exist (Chapters 11 and 14).

Example – electricity retailing. In electricity retailing, a major uncertainty faced relates to the purchase of wholesale electricity on the National Electricity Market. Prices that retailers charge to customers are generally fixed, while wholesale market prices can and do fluctuate. The market price level at any time is uncertain, and so retailers face the prospect of paying more for electricity than they are able to sell it for. This situation can be dealt with through the use of derivative contracts such as options. Options involve paying a premium for the 'right' to buy at a certain fixed price. The premium is paid upfront, regardless of whether the option is exercised.

Example – insurance. The private sector commonly insures its facilities and equipment against damage, theft etc. The insurer accepts risk for a fee.

The public sector commonly insures itself. That is, if a public sector facility is damaged, the loss is accepted.

Example – onerous contracts. Much has been published on so-called 'onerous' contracts. These are contracts where the owner/principal writes the contracts in such a way that the contractor bears the downside, whether or not the contractor is able to influence this downside. The downside is unfairly biased towards the contractor. An example is where the owner does the design but the contractor carries risk associated with design errors. Recommended practice in writing contracts is to have a fair allocation of responsibilities between the owner and the contractor such that the party that can influence a downside is asked to bear the associated responsibility (Chapters 11 and 14).

Example – gambling. When large bets are placed with a bookmaker, the bookmaker may decide to accept the possibility of losing a lot of money, in return for keeping the bet money. Or, the bookmaker will lay off some of the bet money with other bookmakers. Insurance companies may operate similarly where the matter insured is unusual. Re-insurance gives partial or complete coverage from another insurer for a possible outcome on which a policy has already been issued.

Example – insurance. People carry all sorts of insurance – health, vehicle, home, home contents, travel, etc. Insurance is a common way of removing risk. Insurance policies, however, now commonly, include an 'excess' or 'deductible' – claims below this value are not considered; for higher value claims the amount of this excess is payable by the insured and the remainder is paid by the insurance company. Additionally, the insured may lose any 'no-claim bonus' which might take several years to reinstate to its full value. With the presence of the 'excess' and the loss of any 'no-claim bonus', some of the downside is still being carried by the insured, even though an insurance policy exists. With additional premium payment, it may be possible to have all the downside carried by the insurance company. Health insurance covering the gap between a doctor's charged fee and the scheduled fee may not be permissible by legislation. Acceptance occurs when a person decides to 'self-insure'.

Example – mining. The ways in which a mining company reduces risks include the use of joint ventures, insurance, hedging, off-take agreements, capital leasing and contractors.

Example – currency fluctuations. In conducting business between countries, one risk source relates to currency fluctuations. A typical analysis would involve looking at historical records, examining the economic conditions of both countries, and a sensitivity study for fluctuations of plus and minus a few per cent. The evaluation might indicate high, medium or low risk. Responses would include specifying a preferred currency(ies) of payment, some up-front payment, insurance and hedging. Gross movements in currencies would almost certainly not be identified; accordingly, no risk management could be carried out.

Example – crossing a road. Consider an everyday occurrence, namely crossing a street busy with cars. Risk relates to a pedestrian getting injured or even possibly killed. Various responses are adopted by pedestrians: using a pedestrian overbridge/pass or tunnel, looking right-left-then right again (or the opposite in some countries), using a pedestrian (zebra) crossing, the pedestrian getting someone else to cross the road, running, jaywalking, etc. The responses of road authorities/legislators, with respect to pedestrians, include: installing a pedestrian overbridge/pass or tunnel, street lighting, traffic lights, a pedestrian (zebra) crossing, speed-calming devices (for example, road humps), signage on the pavement and/or footpath, vehicle speed restrictions, median strips and pedestrian refuges, fencing separating pedestrians and vehicles, or doing nothing.

Example – oil and gas. In the oil and gas resource industry, uncertainty may be approached through a portfolio approach to projects. In the upstream portion of the business, companies may assign a probability range to the amount of oil that can possibly be contained in a subsurface structure. This probability distribution then has an amount assigned to the Px value (the probability of being larger than this value – Chapter 16) for example. Probability distributions and amounts can be compared and contrasted for a group of opportunities. Execution of this portfolio is by drilling wells to evaluate the multiple projects. For a given budget:

1) A minimum threshold might be set below which the prospect is not drilled – these prospects with high uncertainty may be deferred until more data (that is, wells or seismic testing) are obtained, or only one or two high uncertainty/high impact projects may be chosen to be drilled.
2) Partners may be sought to participate and distribute the exposure to failure (dry hole, cost overruns, etc.). There is less exposure for the operator because capital/expense is shared/diluted and because a premium is commonly paid by partners for the 'privilege of participating'.
3) The amount of partner participation can be varied depending on the position of the prospect in the portfolio. The best projects (moderate to low uncertainty), as defined by the operator, may have limited equity available for potential partners. The operator reduces its exposure to failure, but at the same time reduces its exposure to success by inviting partners to the project. Operator interest may start at 100%, but reduces, for example, to 60% with two 20% partners.

Example – projects. On some projects, risk management practices are adopted in order to account for and eliminate uncertainty. Uncertainty is generally unwanted and hence is responded to. Responses cost money, but eliminating a risk source might be regarded as money well spent. The project operates smoothly with no surprises that may frustrate the project team, surprises which can give rise to potential disputes between team members (resulting in unforeseen delays and costs).

Example – professional indemnity. Consultants and contractors might submit conditional or non-conforming tenders in relation to professional liability indemnity, where the client is attempting to have the consultant/contractor be responsible for acts of professional negligence. Premiums for indemnity insurance are growing and consultants/contractors wish to cap their liability in order to minimize their premiums as much as possible. The premiums, in turn, add to the costs of employing consultants/contractors such that projects start to become not viable. The client, in such circumstances, may take on greater exposure than it might prefer to do to in order for the work to be done.

Example – traffic intersection. At the intersection of two roads, with traffic flowing in both directions on both roads, the volume of traffic and the number of accidents will determine what the road authority does with the intersections. Some choices available to the road authority include: do nothing; install a roundabout; install traffic lights; and separate the traffic by means of an overpass. In the order given, the number of accidents decreases but the cost increases.

Chapter 9

RM summary

This chapter summarizes the essence of risk management as reasoned in the preceding chapters. It is intended to be used as a look-back as to what is covered in the book, as well as an aide memoire for use subsequent to reading the book. Being a summary, the presentation is necessarily terse. The book proper provides the background understanding to this summary. Scale types discussed below are outlined in Chapter 2. Terminology is defined in Chapter 1.

The book applies to all technical and non-technical applications involving risk, not just infrastructure projects.

The book shows that discussion on risk can be reduced to a few simple ideas.

Uncertainty

➤ Risk only exists in the presence of uncertainty; with certainty, there is no risk.
➤ Uncertainty implies probability, likelihood or frequency of occurrence (all used with similar intent), in contrast to determinism (Carmichael, 2013a, 2014). The variables and models are probabilistic.

Inputs and Outputs

➤ Inputs (risk sources) are characterized in terms of pairs {input magnitude; input likelihood}.
➤ Inputs (risk sources) transform to outputs (consequences).
➤ Inputs and their transformations to outputs introduce uncertainty in the outputs.
➤ Outputs are characterized in terms of pairs {output magnitude; output likelihood}.
➤ Particular applications might report output in ways other than as pairs {output magnitude; output likelihood}:
  ■ Either output magnitude or output likelihood is reported for a constant value of the other.
  ■ Expected value of the output, for so-called 'risk-neutral' attitudes.

DOI: 10.1201/9781003343592-10

- Expected utility of the output, where the utility measure is that of the stakeholder for whom the risk study is being carried out – different stakeholders will have different utility functions or curves.
➤ Output magnitude might be defined relative to some desired or base level, rather than an absolute output magnitude.
➤ Quantitative measurement scales or qualitative measurement scales for both input and output magnitude and likelihood can be used. With a qualitative approach, there is always the issue of people's different perceptions of the words, and connotations versus denotations.
➤ Outputs may be upsides as well as downsides.

Risk

➤ Risk = f(Output).
➤ The measurement scale particulars for risk follow from each stakeholder's value system (appropriate to the situation). The pairs {output magnitude; output likelihood} map to risk values peculiar to each stakeholder.
➤ Different attitudes to risk (including what are popularly called 'aversion', 'neutral' and 'seeking' or similar terms) will lead to different measurement scale particulars for risk. Each stakeholder's measurement scale for risk can be different.

Risk Management

➤ Everything about risk management is embodied in the more generic systems engineering (systematic problem solving, systems synthesis).
➤ Risk management should not be considered separate from other management practices. It is usual management but with uncertainty incorporated. All management belongs to synthesis.
➤ Provided that uncertainty is acknowledged by the decision maker, there is no real need to even introduce the term risk, or pretend that risk management is anything special.
➤ The popular approach to risk management is to go through a number of steps, variously described in terms related to:
  - *Definition and Context.*
  - *(Risk Source) Identification.*
  - *Analysis and Evaluation.*
  - *Response.*
➤ Although using different words and numbers of steps, what happens within these steps is no different to that involved in systems engineering (for example, as espoused by Hall, 1962), or as interpreted equivalently in systems synthesis via iterative analysis or systematic problem solving (Carmichael, 2013a), where the steps are given broadly as: Definition; Objective function(s) and constraints statement; Alternatives generation; Analysis and evaluation; and Selection.
➤ Risk management, like practised systematic problem solving, in essence, is trial-and-error optimization.

Definition and Context step

The value system of the relevant stakeholder, and situation, determine:

(i) The measurement scale particulars for risk – based on {output magnitude; output likelihood} pairs – for example {low, medium, high}.

(ii) The objective function(s) by which the feedback adjustments (controls, responses) are selected in the *Response* step – for example, lowest risk, minimum cost.

(iii) The constraints restricting possible adjustments – for example, some adjustments may be unacceptable.

(iv) A model which provides the input-output transformation needed in analysis.

Different stakeholders will select different objective function(s) and constraints, and this will lead to differing adjustments.

The underlying measure of risk adopted has a particular scale based on the value system of the stakeholder, together with the situation.

The presence of the stakeholder's value system introduces subjectivity into risk management. Along with the situation, it also means that risk management is particular to each stakeholder at a point in time and situation, and therefore general statements on risk related to any matter cannot be made unless the value system and situation used are common to many people/organizations. Different stakeholders characterize (including, in some cases, ranking and prioritizing) risk differently.

Risk Source Identification step

(The term identification is used in risk management to mean uncovering or unearthing, and not in the systems modelling sense.) Identifying inputs (sources) requires the same type of thinking as generating ideas/alternatives in problem solving. Some creative thinking is required in this step for all but the most straightforward situations.

This step is strictly risk source identification, not risk identification. *Risk sources* are upstream inputs, while *risks* relate to downstream outputs (Figure 2.1a). Possible outputs are envisaged in the identification step, but only as a means to finding inputs that may produce such outputs. At this risk management step, the output characteristics are unknown, and the practician is attempting to identify inputs.

Magnitudes of inputs might be established through judgement, experience, expertise, education, interviews, idea generation techniques and/or historical or collected data. Likelihoods or frequencies of occurrence of inputs may come from observed frequencies, deductions from mathematical models and/or measures of a person's subjective degree of belief or relative likelihoods.

Both qualitative and quantitative approaches are possible.

If an input (source) is not identified, then risk management related to that input cannot proceed. This is *the Achilles heel of risk management*.

Analysis and Evaluation step

Analysis takes the inputs (magnitude and likelihood) and converts these into outputs (magnitude and likelihood), as in systems analysis generally. The analysis may be carried out qualitatively or quantitatively.

The {output magnitude; output likelihood} pairs are located on the measurement scale for risk appropriate to the stakeholder (as established in the *Definition and Context* step). Each pair could be anticipated to be located or mapped differently for different stakeholders because of their different value systems and situations.

Response step

The *Response* step deals with influencing the inputs (magnitude and likelihood), or transformations from input to output (and hence the magnitude and likelihood of outputs), through actions/decisions, by feedback, in order to address the risks detected. Adjustments (controls, responses) are examined in terms of the earlier stated (*Definition and Context* step) objective function(s) and constraints. Those adjustments which lead to minimizing (or maximizing, as the case may be) the objective function(s), while satisfying any constraints present (admissible values) are preferred. If the steps are done iteratively, the adjustments are examined similarly.

Adjustments are only made if they favourably influence the value taken by the objective function(s). There is also the 'do nothing' alternative, that is, nothing is changed, if applicable. Low or minimal risks (depending on the objective function(s)) may be accepted without further consideration.

Adjustments may occur at different points in time. Two timings of the adjustments may be recognized: immediate responding; and contingency responding.

Adjustment choices might be categorized according to whether they involve (related choices apply for upsides):

- *Elimination* (including: removal, avoidance, transference*); the value of a (minimum) risk objective function goes to zero.
- *Reduction* (including: mitigation, part-transference*, sharing*); this leads to a lower value for a (minimum) risk objective function.
- *Leaving unchanged* (including: acceptance, retention, assuming); the value of a (minimum) risk objective function remains the same.

* As noted earlier, each stakeholder has a different value system; accordingly, for any given situation, the risk to one stakeholder is different to the risk to another stakeholder. That is, risk (as perceived by one stakeholder) is not preserved in any form of transference or sharing (with another stakeholder).

Part B

Applications

Chapter 10

Project delivery methods

10.1 INTRODUCTION

The term 'project delivery method' refers to the way that organizations are contractually and administratively linked on a project. A delivery method (also known as procurement strategy and contract strategy) allocates the different project functions (activities or subprojects such as design, management, construction, …) to the various contracting parties involved in the project (at the organization level); it establishes the roles and interrelationships of the parties, in part establishes the business relationship between the parties, and influences management practices. It can involve issues of timing, but excludes specifics of contractual details (such as conditions of contract including contract payment type). The usage of the term 'delivery method' applies at the organization level and not lower.

Typical project delivery methods will involve an owner, a contractor, subcontractors/trade contractors, consultants of various types and various other parties. Outsourcing or contracting out is implicit in delivery methods. It permits the spread or transfer of responsibilities from the owner to other participants (consultants, contractor, …), usually at a cost to the owner's organization. Outsourcing, whether within a delivery method or elsewhere, can be used as a response within risk management. Outsourcing of parts of a project and the ensuing project end-product may be with respect to finance, design, supply, management, construction (building)/fabrication/manufacturing, operation, maintenance, 'owning' (over different franchise or concession periods or indefinitely) and leasing, and combinations of these.

By adjusting the delivery method makeup, the risk carried by the different parties changes. (Note that it is not a case of transferring risk between parties because risk is not a commodity as argued earlier in the book, but rather it is a case of changing the risk carried by each party by changing the responsibilities of the parties.) It may be, for example, that as the risk to the owner decreases, the cost to the owner increases, but this cannot be stated as a generality, rather it depends on the makeup of the delivery method.

The way the project parties are joined contractually and administratively determines the name given to the delivery method (Carmichael, 2000). For

DOI: 10.1201/9781003343592-12

example, Figures 10.1 and 10.2 show schematically the arrangement of project participants for traditional (construct only) and construction management, respectively.

Irrespective of which delivery method is employed, the same type of people/organizations and skills are employed. The differences between the various delivery methods are one of contractual and administrative links, the way different parties are coordinated and the risk associated with the parties. In simple terms, drawing the methods in the form of block diagrams with connections between boxes being the contractual and administrative links, then each method is a different shuffling of boxes and links between boxes, with the organizations in the boxes 'wearing different hats' (different roles and responsibilities) for each shuffle.

The way the outsourcing occurs (and this defines the delivery method) might be called different names. Table 10.1 shows some conventionally

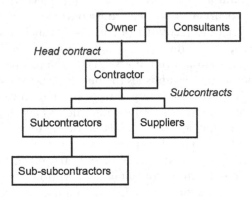

Figure 10.1 Traditional method (showing contractual links).

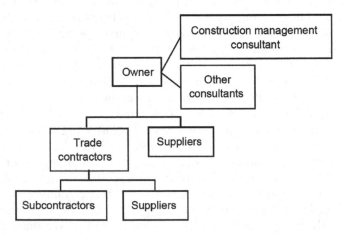

Figure 10.2 Construction management method (showing contractual links).

Table 10.1 Popular names for some delivery methods (Carmichael, 2000)

Item outsourced	Popular name
Construction …	Traditional Construct only Design-bid-build
Design (part), Construction …	Design novation (Novated design-and-construct; Design, novate and construct – DN&C) Detail(ed) design-and-construct(ion) Early contractor involvement (ECI)
Design (part to full), Construction …	Design-and-construct(ion) (D&C) (also termed design-construct, design-and-build, design-build) (Performance contract; Package deal) Design-develop(ment)-construct(ion) (DD&C) Design-document-construct(ion) (Document-and-construct(ion)) Turn-key Managing contractor Design-manage EPC (engineer, procure and construct)
Design and management	EPCM (engineering, procurement and construction management)
Design, Construction, …, Maintenance	Design-construct-maintain (DCM) Design-construct-and-maintain Design-construct-operate-maintain
Management (project)	Project management (Management contracting) Program management (multiple projects) Integrated contract (project management and construction management)
Management (construction, …)	Construction management Professional construction management Construction project management Owner-builder (Management contracting)
Services (other)	… service
Supply	Supply
Design and supply	Design and supply
Many items commonly including finance	Commercial development Concessional methods BOOT, BO, BOT, BOO (build, own, operate, transfer) BOLT, BLO (build, operate, lease, transfer) Public-private partnerships (PPP, P3) Privately funded projects (PFP)

accepted names for some of these delivery methods, but there are many others. Many delivery methods do not have popularly used names, and some are only a minor tweaking or combinations of others but given new names for marketing purposes. Each may be practised with some flexibility, such that there is overlap between them, and hybrids, and the implemented forms may not fit one particular categorization exactly. Thinking up new delivery method names and new shuffling of the boxes seems to be an ongoing pastime in the project industries.

Project industries use a range of delivery methods. Each method has differing characteristics relating to cost management, flexibility, etc.

Placing each of the outsourced items inside boxes such as in Figures 10.1 and 10.2, the links (contractual and administrative) between the boxes may be arranged in many different ways. Each arrangement represents a different delivery method. Accordingly, there are many project delivery methods mentioned in the literature, and new methods (that tinker with existing ideas) arise continually. Subclasses of delivery methods are generated when the contractual and administrative details are tinkered with, for example, by changing the payment type (lump sum, schedule of rates/unit price, cost reimbursable). Much of this tinkering is because of a never-ending pursuit for a panacea – the particular delivery method that has no faults – but of course this particular method does not exist. Often it is a case of 'Everything old is new again' but with a new name.

Public-private partnerships (PPP) and privately funded projects (PFP) can use any delivery method, but commonly use a concessional method of delivery.

Other popular procurement practices that can be incorporated into the delivery method include partnering, alliances, relationship contracting, cooperative contracting, integrated project delivery (IPD), fast-tracking and period contracts.

The choice of delivery method can be fashion driven. Changing delivery methods can be in an attempt at finding a panacea, but no delivery method is fault free, and the same delivery method can perform differently in different situations. Different projects with different owners and circumstances generally mean that there is no one best delivery method for all occasions. Opinion is that a particular delivery method should be tailored to the owner's and project's unique needs. Standard delivery methods may need to be modified. Risk can be one of the determining issues in the selection of a delivery method. But it is only one of many issues.

10.2 BACKGROUND

Many opinions, usually anecdotally based, exist on what are called the 'risk profiles' of delivery methods.

It is commonly stated that one delivery method is 'riskier' or 'contains more risk' than another delivery method. Why people make such unfounded statements locating each delivery method on some risk scale is not known. Such statements are not supportable. Such talk, even though it pervades industry and academia, is meaningless. The parties to the delivery method have risk, not the delivery method.

The meaningless is compounded because the basis for such statements on the 'riskiness' of delivery methods is flawed. Typically, writers take a view of risk based on industry dogma and inflexible assumptions. It is seen that existing approaches in the literature to risk and delivery methods give consideration only to the broad or overview aspects of a delivery method (as described in terms such as in Table 10.1), and largely exclude from discussion the lower-level information within the delivery method, or more importantly the ability to vary this lower-level information. Accordingly, existing thinking is inherently incomplete and misleading.

When selecting a project delivery method, the owner is interested in its risk. (Other parties are interested in their own risks.). The following discussion is in terms of owner's risk.

In spite of pretences to the contrary, no existing treatments are able to establish the owner's risk within a delivery method. The only rational way to build up an understanding of an owner's risk is to examine the structure of a delivery method in a multilevel system fashion. Lower-level information is built up, level by level, in order to understand the owner's risk.

It can be concluded that each delivery method can be manipulated to give a range of risk to the owner, and that no particular delivery method can be said, a priori, to be better or worse from an owner's risk viewpoint, without knowing the specifics of the lower-level information. This conclusion is contrary to popular belief.

It is not possible to say definitively the relationship between a particular delivery method and owner risk without knowing lower-level information particular to each project. Nevertheless, people do say emphatically that a particular delivery method always represents a certain 'risk allocation' between owner, contractor and other parties, when risk can change depending on the lower-level information.

'Risk allocation' between the parties has no meaning, just as it has no meaning within a single contract (Chapter 11).

The performance of a delivery method and the owner's risk not only depend on the contractual and administrative links but also on what is contained in the many contracts linking all the parties, the people within the parties' organizations, and other project-specific issues. The responsibility transfer is done through the appropriate choice of other parties, conditions of contract, including payment type, and other pieces of a chosen delivery method. For example, responsibility for industrial relations issues might be transferred to a contractor. An owner's organization outsources responsibilities to a degree at which it feels comfortable it is prepared to accept the residual responsibility.

10.3 OWNER'S RISK

Owner risk can only be established by looking at lower-level information within a delivery method. Lower-level information that affects owner outputs (outcomes) includes:

- The practices of all the project parties.
- The personnel in the parties.
- The different contracts between the parties including conditions of contract and payment type.

Commonly, the owner transfers responsibilities to other parties until it feels comfortable with the residual responsibilities and possibly ensuing lower level of owner risk. Different sharing, for example, leads to different outcomes (Hosseinian and Carmichael, 2013, 2014a, 2014b, 2016, Hosseinian et al., 2020).

The contracts on the links between parties can be various, while the behaviour of people/organizations within the boxes will differ across projects. The conclusion is that even with the same delivery method, changing the detail of the contracts on the links and changing people/organizations within the boxes will lead to different project outputs and accordingly different owner risk. This is further exacerbated by projects (adopting the same delivery method but) being located in different geographical locations, with different climates, with different access to resources and so on.

That is, each delivery method does not present a constant owner risk across all projects. As well, owner value systems change across projects leading to different scale particulars for risk and different responses for each project.

Extending this argument, it also cannot be said that one project delivery method provides more owner risk than another. By changing the contracts on the links, changing the people/organizations in the boxes, etc., it is possible for any delivery method to provide more or less owner risk than another delivery method.

> No one delivery method can be demonstrated to provide more or less owner risk than another, in spite of what people might believe and promote. The general conclusion is that any delivery method can be

manipulated to provide a range of owner risk, and that no particular delivery method can be said to be better or worse from an owner risk viewpoint, without knowing the specifics of the constituent lower level information.

10.4 DECOMPOSITION OF DELIVERY METHOD

To understand owner risk and delivery methods in the only meaningful way, the lower level details making up the delivery method need to be isolated and addressed.

The dominant lower level details of a delivery method are given in Figure 10.3. Reference to levels here is with respect to Figure 10.3.

Matters associated with different parties, matters affecting the particular connectivity (the contractual and other links between the parties), matters (for example payment types and other conditions of contract) related to the contracts between the various parties, and external influences are the dominant influences on the output associated with any delivery method.

Parties (Level 3). Irrespective of which delivery method is employed, the same type of people/organizations and skills are employed (Figure 10.4).

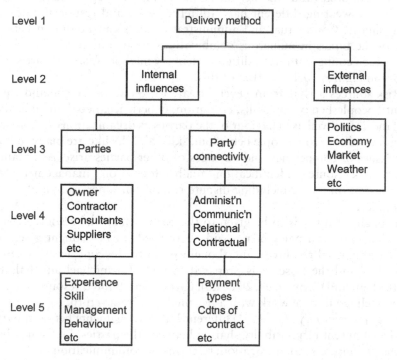

Figure 10.3 Decomposition of a delivery method into its dominant influencing lower levels.

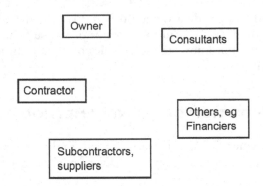

Figure 10.4 Parties (Level 3) involved in the different delivery methods. The differences between the various methods is one of contractual and administrative links (lines) between the various boxes.

The macro differences between the various delivery methods is one of contractual and administrative links, that is the way different parties are coordinated. In simple terms, drawing delivery methods in terms of block diagrams, with connections (drawn as lines) between boxes being the contractual and administrative links, then each delivery method is a different shuffling of boxes and links between boxes, with the people/organizations in the boxes 'wearing different hats' (different roles and responsibilities) for each shuffle. Contractual and administrative links can be drawn between any of the boxes in numerous combinations to give numerous configurations; each configuration is a different delivery method. Examples are shown above in Figures 10.1 and 10.2 of this.

Risk sources arising from Level 5 relate to resources (people and equipment), people behaviour, skills, education, experience, people management and the like as well as behaviour of the parties as organizations. The sources of risk resulting from people behaviour, skills and the like are many in number. Issues of competency and attitudes of other parties arise. As organizations, risk source identification might focus on management and administration issues, including dispute resolution, supervision and quality assurance.

Thought also needs to be given to the parties' financial resources. From the viewpoint of a party being paid, there needs to be thought given to a owner's financial resources, and thought given to the contractor's financial resources and the losses it is prepared to take (Carmichael and Balatbat, 2010; Tran and Carmichael, 2012a,b). Parties with limited resources may be more inclined to take work with a reasonably certain return.

Party connectivity (Level 3). Different link arrangements between parties lead to different responsibility sharing between the parties. Links may be in terms of contracts, administration, relations or communication.

Figures 10.1 and 10.2 show examples of the party boxes linked. Administration and communication links may be different. Figure 10.1

shows the traditional delivery method. Figure 10.2 shows the construction management delivery method. Similar diagrams can be drawn for all other delivery methods (Carmichael, 2000).

Risk sources arising from Level 5 also relate to familiarity between the parties, modes of communication, administration arrangements and contractual issues.

Contracts on the connections (Level 5). Here, a distinction is made between payment types and other contractual conditions, but they could equally well be discussed together.

Contract payment types can either be fixed price (lump sum, schedule of rates/unit price, guaranteed maximum price) or prime costs (cost reimbursable with various fee arrangements), or a combination of fixed price and prime costs parts. Popular usage might describe the contract after the predominant form of payment type used in the contract, although different payment types might be present in the one contract. As well, there are convertible contracts which begin on the basis of one payment type and convert to another payment type at a defined point in the project (Carmichael and Karantonis, 2015).

A general belief is that the owner should, wherever possible, define the work as carefully as possible beforehand, irrespective of the contract payment type used. This could be anticipated to reduce the uncertainty to both parties. The financial concerns decrease for the owner (increase for the contractor) in the following order: lump sum; lump sum (no escalation/rise and fall; no delays except owner-caused); lump sum (no delays except owner-caused); lump sum (no escalation/rise and fall); lump sum (with escalation/rise and fall; with delays); lump sum (with escalation/rise and fall; with compensation for delays); lump sum (with many adjustments); schedule of rates; cost-plus fixed fee; cost-plus percentage fee.

As more financial output (outcome) responsibility is taken by the owner, it could be anticipated that the contract price would decrease. There may be an overall saving to the owner by accepting some responsibility. Generally, there is an expectation from owners that contractors should bear some financial output responsibility, though some owners take this to an extreme and (through choice of conditions of contract, including payment type) ask contractors to bear essentially all financial output responsibility. Owners with limited financial resources may not wish to take the financial output responsibility associated with a cost-reimbursement contract, even though it may be more suited to the type of project.

For a lump sum contract, the contractor stands or falls based on its performance relative to the lump sum price that it bid; the responsibility associated with completing the work at the tendered price, in the tendered timescale and to the required standards, lies with the contractor. However, there may be provision for adjustments. The contractor's tender should accordingly reflect this, though in a competitive tendering environment, contractors may (perhaps unwisely) elect not to allow for some uncertainty for fear of being non-competitive. Tenderers accept the situation that there may be more involved in the work than they have allowed for in their tender, although a lot of such unknowns can be removed through the inclusion of special contract clauses, for example covering 'rise and fall', or guaranteeing quantities. Most of the uncertainty associated with the work (for example, related to weather, industrial relations and external matters) might be carried by the contractor; some might not be under the contractor's ability to influence.

In a schedule of rates contract, there is a downside to the owner if its estimates of quantities are wrong or un-anticipated work arises. A schedule of rates contract represents a shared financial output responsibility between the parties. The contractor may bear the responsibility associated with inclement weather, industrial relations and other external matters. Unbalanced bids are another issue.

Under a guaranteed maximum price contract, all financial output responsibility is being transferred to the contractor, and its price should reflect this.

Cost plus arrangements may be used where the project has features making the work involved difficult to appraise. Of all the contract payment types, most financial output responsibility is with the owner; financial output responsibility carried by the contractor under other payment types is transferred to the owner (for example due to industrial relations matters, weather, supplier delays, …). Under a cost plus a fixed percentage fee payment, cost plus a fixed fee payment and cost plus a variable percentage fee payment, most of the financial output responsibility rests with the owner. In cost plus with bonus/penalty or target contracts, there can be a kind of contradiction. When a risk source materializes which is other than something for which due allowance has been made in the target, or when an operation or an amount of work has to be done which was not included in the target, the target must be changed to be fair to the contractor. Contracts, that involve extensive modifications to the target as the work proceeds, can be a source of friction and dispute. Under a gainshare/painshare arrangement, the financial output exposure carried by parties may be capped.

With a guaranteed maximum cost contract, where the contractor is responsible for everything beyond a stated maximum ('upset') cost of the work, such financial output responsibility is accordingly reflected in the size of the fee (a fixed amount). The more indefinite the scope (of work) is at the start, the higher will be the fee.

Of the prime cost versions, a guaranteed maximum cost fee-based approach passes most financial output responsibility to the contractor. The contractor may even be asked to bear responsibility for matters that are not within its influence.

Other contractual conditions (Level 5). Conditions of contract can be written in many ways, ranging from being considered 'fair' to both parties, to being 'onerous' on one party at the expense of the other. There is a view that conditions of contract should be written such the party which is best able to deal with something is allocated that responsibility but implementation of this philosophy through conditions of contract is difficult; there may not be a direct relationship between the words in the conditions of contract and a downstream exposure. Chapters 11 and 14 discuss this more fully.

Particularly relevant bits of conditions of contract that affect project outputs include: the nature of the work, design responsibility, nomination, payment provisions, retentions, cost adjustment, equipment, variations, approvals, extension of time, claims, delays (Carmichael, 2009), liquidated damages, the program and insurance. But this list is by no means complete.

In addition, there are the other contract documents such as the specification and bill of quantities which affect outputs.

External influences (Level 2). External influences refer to anything that impinges on the delivery method but which is not specifically included in the constituent parts of the delivery method. External influences include, where not mentioned above, industrial relations matters (external to the project), the market and the economy.

10.5 OWNER RISK DEVELOPMENT

Owner risk associated with a particular delivery method can only be established from the given particulars proposed to be used in the delivery method's lower levels. It is not possible to say in any general terms that one particular delivery method (as described in terms such as in Table 10.1) represents more or less owner risk than any other delivery method.

The magnitude and likelihood of any output relevant to the owner can only be established by information from one level to the next higher level, and doing this from the lowest level in Figure 10.3 up to the highest level. And this cannot be done to give a general result, but rather only a result peculiar to a particular project. (A different risk conclusion would be obtained if done from the point of view of a party other than the owner.)

Typically, project output will be in terms of cost and/or time overruns, but other outputs might be considered. Risk follows knowing the output.

In looking at the impact of the delivery method lower levels on the project output, a qualitative approach may be all that is possible. Data may not be available in practice to do any accurate quantitative analysis. In most cases judgement and opinion will be all that is available.

Example. At all levels of Figure 10.3, there are ranges of commonly occurring existing practices, which collectively produce ranges in project output uncertainty, and project output magnitudes. This leads to a range of owner risks. Consider, for some typical lower-level matters, the following ranges of practices:

Contract payment types:
Guaranteed maximum price; cost plus a percentage fee.

Conditions of contract (examples):
   No escalation clause; escalation clause present.
   No latent conditions clause; latent conditions clause present.
Party experience and/or skills:
   Extensive experience/skills; no experience/skills.
Party organization management:
   Appropriate management style; inappropriate style.
Party organization administration:
   Systematic, documented processes; ad hoc processes.
Communication:
   Timely, appropriate communication; grape vine.
Relational:
   Parties work well together; dysfunctional relationship.
Economy:
   Readily available skills/labour; tight employment market.

This will result in each delivery method having a possible range of owner risk from low to high. Of course, prudent owners adopt the risk level they desire by appropriate choice of delivery method lower level detail, but the conclusion remains that no one delivery method can be said to have more/less owner risk than another without knowing the lower-level information, and owner risk associated with delivery methods can only be meaningfully discussed in terms of the lower level details, and not at the macro delivery method level.

10.6 CLOSURE

Risk management is particular to each stakeholder (contracting party) at a point in time and situation, and therefore general statements on risk related to any matter cannot be made unless the value system and situation used are common to many people/organizations. Thus, it cannot be said in general terms that one project delivery method or conditions of contract contains more/less owner risk than another, even though such claims are prevalent in the literature (Carmichael, 2000). Different people/organizations character-ize (including, in some cases, ranking and prioritizing) risk differently.

The general conclusion is that each delivery method can be manipulated to give a range of owner risk, and that no particular delivery method (as described in terms such as in Table 10.1) can be said to be better or worse from an owner risk viewpoint, without knowing the specifics of the constituent lower levels. The converse of this is that prudent risk management practices on behalf of the owner can manipulate its risk to whatever is desired for any delivery method; various responses applied to the delivery method lower levels might be tried.

Given a particular situation, it is possible to establish owner risk associated with any delivery method, but that this owner risk will be different in different situations and for different owners. Any delivery method can lead to a wide range of owner risks.

These conclusions are contrary to popular belief.

Chapter 11

Risk allocation in contracts

11.1 GENERAL PRINCIPLES

A contract is an enforceable agreement between two parties. An engineering-style contract typically involves general conditions of contract, special conditions of contract, a specification, drawings, perhaps a schedule/bill of quantities and other documents. For definiteness here, the two parties to the contract are referred to as owner and contractor, but the below comments also apply to consultants and subcontractors and equivalent terminology such as builder and principal.

The general conditions of contract might be a standard document such as those authored by professional bodies including standards associations, or bespoke for the particular project owner. With the owner largely the one able to influence the clauses within the general conditions of contract, the clauses might be adjusted by the owner or owner's legal representative in the belief that the risk to the owner is decreased and the risk to the contractor is increased.

This might be referred to as 'risk allocation'. The Abrahamson Principles are commonly quoted (Aust Constrn Law Newsletter, Issue 14, p. 8, 1990) and provide:

> 'The basic principles of allocating obligations and/or risks for all projects … are those expounded by international construction lawyer Max Abrahamson. Those principles dictate that a party to a contract should bear a risk where:
>
> - The risk is within the party's control.
> - The party can transfer the risk … and it is most economically beneficial to deal with the risk in this fashion.
> - The preponderant economic benefit of controlling the risk lies with the party in question.
> - To place the risk upon the party in question is in the interests of efficiency, including planning, incentive and innovation.

DOI: 10.1201/9781003343592-13

- If the risk eventuates, the loss falls on that party in the first instance and it is not practicable, or there is no reason under the above principles to cause expense and uncertainty by attempting to transfer the loss to another'.

This then gets implemented assuming: risk to one party is measured the same as to the other party; each party knows the risk of the other party; and that each party knows the value system of the other party and hence the objective function and constraints and the responses preferred by the other party. And clearly, none of this is the case.

Any argument used for risk allocation, and risk sharing generally, does not hold if the proper meaning for risk is used. (Risk allocation thinking such as in the above principles was developed based on a range of lay meanings for risk.) Reducing an owner's risk does not translate to the contractor picking up the equivalent amount of risk.

Risk is not transferable between organisations (or people) because each organisation has a different value system, defines risk differently, and implements responses differently. Risk interpretation is peculiar to each person and organisation.

It may be that one party's risk is reduced by some action, and the other party's risk increased by the same action; however, one party's increase is not the negative of the other party's decrease; the increase and decrease in risk are not related because different risk measurement scale particulars and value systems are being used by each party. It is not like a zero-sum game (Chapter 8 and Carmichael, 2013a, 2020b).

Rather than talking of risk allocation in contracts, it would be more appropriate to talk of allocating responsibilities, to which each contracting party can establish its associated risks. Each party can never know what the risk of the other party is because the value system of the other party (and on which the risk measurement scale particulars of that party are established) is private. With changes in responsibility between parties to a contract, it does not follow that the change in risk to one party is matched by an equal (but opposite) change in risk of the other party. Each party can never know what the preferred response of the other party is because the value system of the other party (and which establishes that party's objective function(s) and constraints) is private. The preferred response of one party in general is not the same as the preferred response that the other party might use, and hence the associated outputs (magnitude and likelihood) of each party will be different, and hence the risk to each party will be different.

On the surface, the Abrahamson's principles sound sensible, but they cannot be implemented at the conditions of contract level alone. While the spirit of the Abrahamson Principles is worth supporting, they need revisiting and rewording using a proper understanding of risk. Something along the following lines is suggested where the term risk has been removed completely.

Responsibility should be allocated to the party which is believed better able to deal with that responsibility, given that uncertainty is present.

This is not saying that it is known that one party is better able to deal with a responsibility, rather it is believed better able to deal with a responsibility.

Contract clauses are notoriously known for being open to potentially more than one interpretation. Case history, judge opinion and English meanings, along with extrinsic matters such as past usual practice and legislation can influence the interpretation of contract clauses. Contract clauses are then superimposed on a project – the scope (of work) defined by the contract – but the way a project is carried out is generally not prescribed in the contract. Accordingly, it can be difficult to say how any particular contract clause might establish the owner's and contractor's risks, even if the complete exclusion of a responsibility to one of the parties is entertained.

There is also the added issue that industry undesirably blurs the distinction between a contract and a project. Many, particularly larger contractors, wrongly refer to a project as a 'contract'. And so, discussion on 'risk allocation' in a contract blurs over to include 'risk allocation' in the project.

11.2 CONTRACT PAYMENT TYPES

A contractor gets paid for work done or services provided. This payment may follow a number of forms including the following:

Fixed price – guaranteed maximum price (GMP)
Fixed price – lump sum
  • No escalation/rise and fall; no delays except owner-caused.
  • No delays except owner-caused.
  • No escalation/rise and fall.
  • With escalation/rise and fall; with delays.
  • With escalation/rise and fall; with compensation for delays.
Fixed price – schedule of rates, unit price
Prime cost – cost-plus, cost reimbursable
  • Fixed fee.
  • Percentage fee.
  • Others.

As the above list ascends, the contractor's potential for loss generally increases, assuming that the contractor's estimating and tendering is appropriately done pre-contract, and the same work method and productivity is used by the contractor. Some literature therefore suggests that as the list ascends, the contractor's financial output risk increases. However, the unknowns in this are the likelihoods associated with the various contractor losses, and hence no relative statement can be made on the contractor's risk for each payment type.

Some literature suggests that on ascending the above list, the owner is transferring financial output risk to the contractor. As with the earlier comments in Chapter 8, such a statement is unsupportable because risk to the owner and risk to the contractor are measured differently as a result of the different value systems of the owner and contractor. From the owner's point of view, all that can be said is that uncertainty about the final cost of the work becomes lower on ascending the list. In general, this uncertainty will not be the same as that surrounding the contractor's losses. Also, it could be anticipated that the owner's cost to have the work done will increase on ascending the list. And so, there may be an overall saving to the owner by accepting some uncertainty in the total cost, equivalently accepting some risk.

Public accountability may inhibit public sector bodies from using anything other than fixed price contracts. Owners with limited financial resources may not like the uncertainty in the final cost associated with cost reimbursement contracts, even though they may be more suited to the type of project.

From the contractor's viewpoint, thought is given to the contractor's financial resources and the potential for losses that it is prepared to take. Contractors with limited resources may be more inclined to take work with a reasonably certain return (Carmichael and Balatbat, 2010; Tran and Carmichael, 2012a,b).

> It is not possible to say definitely the relationship between a particular payment type and risk, even though such statements appear in the literature. The conditions of contract, among other project variables, also need to be examined.

The above looks at financial matters but, of course, there are other matters influencing the owner's choice of payment type.

11.3 CLOSURE

Risk management is particular to each stakeholder at a point in time and situation, and therefore general statements on risk related to any matter cannot be made unless the value system and situation used are common to many people/organizations. Thus, it cannot be said in general terms that increasing (decreasing) one party's risk will lead to an equal and opposite decreasing (increasing) of risk to the other party, even though such claims are prevalent in the literature (Carmichael, 2000). Different people/organizations characterize (including, in some cases, ranking and prioritizing) risk differently.

Chapter 12

Workplace health and safety

12.1 STATE OF PRACTICE

The treatment of risk within the workplace health and safety discipline has developed in its own way because of the usual requirement to eliminate injuries and death rather than generally manage them. In some ways, health and safety risk management is a special case of risk management covered in Part A. This chapter provides commentary on the practice of risk management within the workplace health and safety discipline.

> In the literature, applications concerning risk in health and safety in the workplace appear fortunate in that they are able to confuse the difference between hazard (input), consequence (output) and risk without seemingly incurring any serious issues in terms of achieving desired health and safety results.

Strictly, the hazard is the risk source (input) and it leads to some injury or death (output).

> The hazard is not the risk; injury or death is not the risk; risk relates to the magnitude and likelihood of injury or death.

The sloppy and wrong interchangeable use of 'hazard', 'consequence' and 'risk' in the health and safety literature seems to cause few anxieties among practitioners because:

* The likelihood of a hazard (input) is usually assumed to translate directly to the likelihood of an injury or death (output); a one-to-one transformation is usually assumed between input (risk source – hazard) and output (consequence – injury, death). In most cases, the path from hazard to injury is straightforwardly apparent.
* Typically, because injury and death are involved, the intent is to eliminate all hazards (inputs) which lead to serious injury or death (outputs); very minor injuries might be tolerated if of low likelihood and

DOI: 10.1201/9781003343592-14

hazard removal is difficult. By contrast, in non-health and safety applications, broad risk management occurs meaning that a value system of the stakeholder is in place and some residual risk may still exist after the response step; in health and safety applications a value system is not required to be stated because the preferred response is always one that leads to output elimination.

- Risks are measured from the point of view of a universal worker – all workers are assumed to have the same value system (generally not required) and measure risk in identical ways; there is only one stakeholder from a risk management viewpoint and this is this universal worker.
- Risk is always measured from the worker's viewpoint, not from the point of view of their organizations or managers within those organizations. The undertaking of risk management by an organization as pre-emptive case building should a future worker injury occur also takes the same viewpoint.
- The steps in risk management are typically carried out in a very mechanical checklist and tabulated way.
- Practices in hazard identification and the implementation of responses are frequently carried forward from one project to the next without thinking too deeply about any changes that might be present in going between projects.

And so, health and safety publications freely and loosely interchange the use of the terms hazard, consequence and risk, talk wrongly of 'risk identification', talk wrongly of 'risk control' and so on. This demonstrates a lack of deep understanding of risk. Rarely do health and safety publications define what they mean by risk. This leads to the use of the term 'risk' to ambiguously mean more than one thing (including hazard, consequence, risk) and hence does not provide the certainty necessary, particularly if referred to in a legal document or standard/code of practice.

12.2 EXAMPLES

Some health and safety publications are better than others, but they all exhibit deficiencies of some form. For example, the document Construction Work Code of Practice (Safe Work Australia, 2018) has a focus on workplace risk, but nowhere does it define risk.

> '[It provides guidance] ... on how to eliminate, or if that is not possible, minimize the risks relating to construction work. [It] ... sets out a list of construction work that is considered to be high risk for the purposes of ... health and safety regulations. It [high risk construction work] is construction work for which a safe work method statement (SWMS) is required'.
> (Safe Work Australia, 2018, p. 8)

The document does not define 'high risk', but based on the examples given, it would relate to serious injury or death from activities that could be anticipated to occur on construction sites – that is, outputs or consequences with high magnitude and non-negligible likelihood.

Guidance in this Code for managing risks is given through risk management steps as adopted by lay persons. (In the health and safety literature, the term 'control' may be favoured over the term 'response'.) A hierarchy of controls is suggested (Safe Work Australia, 2018, p. 12):

'You must, so far as is reasonably practicable:

- First, eliminate risks by eliminating hazards; this is the most effective control measure.
- Then substitute hazards with something safer, isolate hazards from people and/or use engineering controls to minimize any risks that have not been eliminated.
- Then use administrative controls to minimize any remaining risks.
- Then use personal protective equipment (PPE) to minimize any risks that remain'.

A Safe Work Method Statement (SWMS) is a tabulated and structured approach to risk management. The risk management steps outlined in Part A underly what is in a SWMS. Typically, the columns in a SWMS comprise (Safe Work Australia, 2018): List of work tasks; Hazards and risks; and Control measures. However, more appropriate columns would be: List of work tasks; Hazards (inputs); Consequences (outputs); Likelihood; Risk; and Control measures.

Health and safety legislation is growing, and in doing so naturally incorporates the term risk. Unfortunately, in developing health and safety legislation, the term risk is being used in a lay or dictionary sense (Chapter 3) possibly to suit the non-technical backgrounds of lawyers, and hence the legislation translates ambiguously even if the intent and purpose of the legislation is understood.

12.3 INJURIES STILL OCCUR

It would seem that with workplace risk management practices in place that no injuries should occur or should occur very rarely. Unfortunately, because of intimate human involvement in work practices, the Achilles Heel of risk management (Chapter 6) is exampled often.

Example. An open trench is identified as a hazard. A barrier and warning signs are erected to stop people falling in the trench. The risk management is complete. Human nature then intervenes and a person ignores the warning signs and barrier and climbs over the barrier. An injury follows. The risk source of someone ignoring their own health and safety has arisen but not catered for in the original risk management.

There are many other similar examples where people have ignored their own health and safety, their health and safety training, and safety devices installed in the workplace. None of these activities of people ignoring their own health and safety, their health and safety training, and safety devices installed in the workplace get listed as risk sources and hence associated risk management is not carried out. Injuries follow.

It would seem that, with health and safety and people involvement, risk management can never be totally complete. People are unpredictable in their actions giving rise to unidentified risk sources.

Chapter 13

Management and project management

13.1 INTRODUCTION

The chapter provides commentary on the practice of risk management within the management and project management disciplines.

There is general confusion across the management and project management literature as to what risk is and hence what risk management entails. It appears that the further a discipline is from being a technical discipline, the looser the use of terms becomes.

The general management literature's use of terminology is possibly the least rigorous of any discipline. This is possibly due to the young nature of the discipline compared to established technical disciplines, and to writers without science or engineering backgrounds. For related commentary, see for example Carmichael (1996, 1997a, 1998, 2019). The project management literature tends to be a bit more quantitative and a bit more rigorous.

> A wide range of meanings for the term risk can be found in the management literature. All lay meanings given in Chapter 3 are used confusingly throughout the management literature.

As given in Carmichael (2013a) and in Chapter 2, management is an example of systems synthesis (systems engineering, systematic problem solving). And risk management is management where uncertainty has been incorporated. Accordingly, treating risk management separate from management, as occurs in the literature, indicates a lack of understanding of the inverse systems nature of management and risk management.

Management-style terms are used across different disciplines, and this is something that is not encountered, for example, by technical terms in engineering, medicine, architecture and other technical disciplines. The development of management is relatively new and it is still finding its way. The only way for management to progress is for people to agree on a set of definitions of all management terms. As long as each word conveys multiple meanings, management will not progress. The verbal models of management will be of little use. It will continue to allow charlatans to promote themselves as

DOI: 10.1201/9781003343592-15

managerial experts, and in organizations, it will continue to entrench 'managers' who do not know what management is, but who usually know the jargon, and none can ever be proven wrong.

The project management literature is equally confused; risk is many things, mixes many ideas and lacks consistency even within single documents. This puts a cloud over how risk is then managed, and the understanding of what risk management is. All lay meanings given in Chapter 3 are used confusingly throughout the project management literature. Superficially, the risk coverage given in the project management literature looks appealing, but on closer inspection there is confusion.

Projects are comprised of stages (phases), and risk management can be applied at all stages. Irrespective of when risk management is considered within a project's lifetime, the approach given in Part A applies. Each stage of a project, when addressing risk specific to that stage, may also look ahead to any potential impact on subsequent stages of the project.

There is a need for the management and project management literature to become more rigorous in its dealing with risk management. The approach of Part A would provide this.

13.2 EXAMPLES

Two examples are given here, and are described in terms that fit Part A rather than a loose approach to risk management.

Example – Construction of a building. Delays during a project affect the completion time of the project and may bring about contractual breach. The delays may be brought about by poorly trained workers and poor project management. Poor workmanship may not be evident throughout a project but the results of poor workmanship may only become apparent during the post-project (asset) phase. Risk relates to delays and repair, while the risk sources relate to worker skills and worker management.

Example – Fast-track transmission substation building projects. These projects compress the project stages and overlap them to some extent in order to complete the project within a restricted time frame. There is very limited design time for the civil/structural engineers, and construction commences prior to final completion of electrical and earthing design. Their design can often affect the location of concrete columns, require large conduit holes in building walls and require further structural support for equipment, and external pavement re-design. These changes inflict significant time delays on project stages and extra costs. The implication of fast-track projects, is that they are more susceptible to uncertain events and their impact accumulating into the following stages of a project. Risk management involves continuously looking at possible risk sources. Risk relates to extra cost and delays.

Carmichael (2016a) covers other deficiencies in the way risk is dealt with within the project management and management literature.

Chapter 14

Tenders and contracts

14.1 INTRODUCTION

The following is written in terms of an owner-contractor relationship, but can also apply to contractor-subcontractor relationships and other contractual relationships that may exist on a project.

At the time of tender preparation by the contractor, there are a number of matters that are examined. These matters range over administrative, contractual, financial, time and the work itself. Identification, and analysis and evaluation within risk management practices are carried out before a tender is submitted. It is recognized that this is prudent behaviour, but not universally carried out.

The chapter includes notes on tender examination and tender checklists. Chapters 10 and 11 provide additional insight on 'risk allocation', contract payment types and delivery methods. The notes on tender examination identify and discuss some of the more important issues facing a contractor preparing a tender. The notes on tender checklists and Chapters 10, 11 and 14 provide ways of dealing with tender and contract risk issues.

14.2 TENDER EXAMINATION

Before spreading business overheads, contingencies and profit over the direct and indirect cost items, the tender documents are studied closely by the contractor, particularly contractual matters related to payment or work quantity and quality matters. Some tenderers do not do this, but it is considered to be poor practice not to do so.

It is generally argued that the tenderer needs to know the owner and its requirements. There may be a need to know such matters related to: the fairness of the tender requirements; the method and ethical practices to be adopted in the evaluation of tenders; the ability to qualify a tender and price any qualification; and the financial standing of the owner.

Separate from owner issues, but related to the tender itself, there is a need to know matters such as: the cost of tendering; the tenderer's financial status

DOI: 10.1201/9781003343592-16

and current and future workload and the type of work; particular provisions in the tender documents especially related to the treatment of rise and fall and payments; confidentiality of tenders; disclaimers in the tender documents; conflicts among the tender documents; that relating to nominated subcontractors; the type of contract; payment and time provisions in the contract; and the status of the bill of quantities. These are considered in the following paragraphs.

The cost of tendering. Contractors typically carry the cost burden of preparing their tenders; only on some projects may owners reimburse contractors. This preparation cost is an overhead that becomes a part of the tender price. Given that contractors may only win say one in five tenders that they submit, many would argue that the cost of tendering should be borne by the owner as it is the owner who is getting the benefit from competitive tendering. In the alternative, and to prevent an unnecessary cost to industry, owners can adopt prequalification and keep the number of invited tenders small.

There are two sides, of course, to the debate on whether an owner should directly bear the costs of tender preparation associated with that owner's project or whether the cost of tendering for all projects should be included in the contractors' overheads and smeared across those projects that result from successful tenders.

Owners usually take the following view:

- The preparation of tenders is an integral part of a contractor's operations and hence should be treated similarly to other operating costs. The cost is considered a reasonable impost better met by contractors.
- With the tendering cost borne by owners, it is argued that inefficient tenderers would be reimbursed inappropriately.
- This debate is only relevant for the larger projects which are undertaken by the larger contractors. The larger contractors have their own estimating/tendering departments or sections and the costs of these are included in the contractors' overheads.
- With the owner paying for tendering costs, it is argued that this amounts to 'double dipping' and contractors are seeking 'unjust enrichment'.
- With two-stage tendering, the owner may contribute to the cost of tendering for the reduced number of tenderers in the second stage.

Contractors, on the other hand, take a different view:

- There are projects and/or conditions of tendering that impose exceptional costs on the tenderers, costs in excess of those associated with tendering on conventional projects with conventional conditions of tendering. The owners who create such costs should directly bear those costs rather than spread them to other owners (via contractor overheads).

- On large or difficult projects, only larger contractors can afford to tender. If the owner pays the cost of tendering, this permits the entry of smaller contractors into the competition. It may also permit smaller contractors to become larger.
- Where bills of quantities are not provided or, if provided, are not guaranteed, this increases the cost of tender preparation. All tenderers effectively have to develop their own bills of quantities, duplicating effort and cost.
- Where the owner uses ideas and methods from submitted tenders to request again tenders based on the 'borrowed' ideas and methods, the tenderers have the cost of tendering twice, while the owner has the benefit of the tenderers' expertise without compensating the tenderers accordingly.
- Where tenderers are involved in design, develop and construct work, design and construct work, concessional delivery work or other involved delivery methods, both owners and contractors recognize a case for reimbursing contractors their tendering costs.
- On 'high tech' projects involving considerable expertise and special construction techniques, both owners and contractors recognize a case for reimbursing contractors their tendering costs. The payment by owners to contractors may be called euphemistically 'development grants' or similar but is intended to assist with tendering costs in order that the owner can get truly realistic competition between the selected tenderers.

Financial status. Some projects fail because of inadequate financial resources of the contractor – an inadequate ability to finance the work. Sometimes this may be due to an inability to handle an un-anticipated loss, or a cash flow determined by restricted material and labour availability.

There is also the possibility of owner insolvency.

Complimentary to an examination of the financial resources of the tenderer is an examination of the owner's past performance for the payment of progress claims including variations. Does the owner (or its representative) have a record for prompt certification of work done and prompt payment of the associated claim? What is the credit rating of the owner? Does the owner have a good record with respect to the settlement of accounts? (Carmichael and Balatbat, 2010; Tran and Carmichael, 2012a,b, 2013).

Current and future workload. Consideration is given to not overcommitting plant and personnel in the one extreme and in the other extreme providing sufficiency of work. High utilization of plant and personnel lowers overheads. Insufficient work can act as a demotivator for staff. The provision of continuous work for subcontractors and suppliers ensures good ongoing relations with the contractor.

Large contractors could be anticipated to be able to cope with fluctuations in workload, better than small contractors.

A certain amount of continuous work is necessary for any contractor to stay in business.

Type of work. Large contractors would more likely have access to better records of costs and productivity, on which to base their bids, than small contractors.

Rise and fall, escalation. Ideally, for a project over an extended period (and certainly over more than one year), the contract conditions should contain price adjustment mechanisms to account for changes in the cost of labour and materials. Where such a mechanism does not exist, the tender can make an allowance for this. Not all materials are purchased at the start of the project.

Confidentiality. Information provided in a tender needs protecting. Confidentiality might be obtained by including in the tender a clause requiring the owner to protect the confidentiality of the tender. Additionally, a tender might contain intellectual property held by the tenderer.

Disclaimers. Invitations to tender may contain disclaimers, for example, the documents: may transfer all responsibility from the owner to the tenderer for information contained in the tender documents; require the tenderer to have satisfied itself about the site conditions and the work; require the tenderer to make its own investigations; require the contractor to repair owner-supplied equipment and materials; or may not allow any claim against the owner because of something the owner does or neglects to do. Some disclaimers may or may not turn out to have no legal effect, depending on the situation.

Information, such as soil conditions, might be made available by the owner but not guaranteed by the owner. Some information provided to tenderers may not be accurate.

Although disclaimer clauses are generally written by lawyers on behalf of the owner, they are not written with a full working knowledge of how the project management industry conducts its business. Due to the interpretation of such clauses by both parties, contractual disputes may arise.

For a contractor to accept such clauses in a contract is flirting with danger and in most cases the pricing structure of the tender will take into account the implications of such clauses on the overall project, or the contractor will qualify its tender based on a fair and reasonable allocation of responsibilities. Hence, the likelihood of non-conforming tenders is high. The owner on the other hand might believe that all responsibilities should be passed onto the contractor, which puts the owner-contractor relationship on the back foot from the beginning.

Conflict is believed best avoided if a fair and reasonable approach to responsibility allocation is adopted during the tender stage.

As an example, on one project the contract was tightly written and onerous on the contractor. Initially everything was fine, but as events began to deviate off the predicted path, the contractor was placed under significant financial pressure. As a result, the project mood became very hostile. Even though contractually the owner had the high ground, the reality was that a project dispute would end up in court and the court outcome would be unclear. The owner decided to alleviate some of the issues by granting some variations that the contractor may have been morally entitled to, if not contractually. This concession was considerably less than any legal costs if the other path had been chosen.

A contract sets out the parties' rights and obligations. If one party's obligations are too harsh, it will result in conflict (a negative output) and continual reference to the contract, leading to a loss situation for both parties.

Conflicts among the documents. Tender documents are not always clear, unequivocal and complete. Engineering-style contracts consist of a number of documents, and these can contain inconsistencies, ambiguities, omissions and errors. Inconsistencies, ambiguities, omissions and errors would usually represent exposure to the owner. Precedence of documents might be nominated in the tender documents in an attempt to remove any inconsistency.

Tender documents can also be inadequate for their purpose intended, namely for enabling the work to be priced and carried out properly.

Nomination. Subcontractors might prefer an arrangement where they are nominated by the owner to do their respective trade work. The reason for this is an increased certainty over payments, should the contractor default. However, contractors may not like such an arrangement, preferring to work with their known subcontractors.

Subcontractors can also be subject to pressure from a contractor to reduce their prices (bid shopping). This may win the subcontractor the work but to the subcontractor's disadvantage. There is also bid peddling.

Contract type and delivery method. Most of the notes in this chapter are written around the traditional contract delivery method (Carmichael, 2000, 2002). When tendering under other delivery methods there are some particular points of note. Even within a traditional delivery method there may be departures from standard practice and all non-standard requirements need attention in the tender examination.

Design and construct delivery requires more work in preparing a tender than traditional delivery because of the need to also prepare a design. Accordingly, the cost of tendering is higher, but this may be offset partly by being paid a design fee. Confidentiality of the contractor's design needs to be ensured in order that the owner does not use the contractor's design and get construct-only prices from competitor contractors. Such prices, based on construct-only, could be anticipated to be lower than design-and-construct prices.

Commonly, design and construct work tendering is initiated by a contractor with third-party design consultants being used. There is always the issue of the design work and design time being to no avail should the tender be unsuccessful. Is this a reasonable cost for the design consultants to bear or should the contractor contribute some funds towards the design? If the total tender is prepared by one design and construct organization alone, then this issue does not arise though there is still, of course, the tender preparation cost to be borne.

Different contract payment types and delivery methods involve different exposures for the owner and tenderer.

All else being equal, it is usually assumed that financial outcome exposure to the owner decreases on going from cost-plus to schedule of rates (unit price) to lump sum payment types. However, this exposure can be changed, for example, by varying the conditions of contract, other contract documents, project players, work type and external influences (Chapters 10 and 11).

The owner's risk in each delivery method (Table 10.1) can be varied to whatever level wanted by varying, for example, the conditions of contract, other contract documents and payment type associated with each contract as well as varying the project players, work type and external influences. Alternatively, by appropriately selecting conditions of contract, other contract documents and payment type associated with each contract as well as project players, work type and external influences, the owner gives itself the desired risk that it is prepared to accept (Chapters 10 and 11).

Payment provisions. Progress payments released within a defined period of the completion of tasks are preferred. Payments to subcontractors based on the contractor first being paid may not be legal. Careful scrutiny is given to the payment provisions within the tender documents including who is the owner, when will payments be made, what certification or approval is necessary in order to be paid, and under what conditions can payments be withheld. Nominated subcontractor payments are discussed above. In some jurisdictions, there is legislation dealing with security of payments to contractors and subcontractors.

From the owner's viewpoint, insolvency of the contractor because of non-payment is to be avoided. There is also the possibility of owner insolvency.

Time provisions. Time provisions in a contract ideally reflect the availability of labour and materials. Specifications may be for components that are in difficult supply.

Bills of quantities. There is a debate in the industry whether owners should provide a bill of quantities (BOQ), or if a bill is provided whether the owner should guarantee the information contained within it.

At one extreme, favoured by contractors, is that there be a complete, highly detailed and measured bill fully guaranteed by the owner. At the other extreme, pushed by some owners, is to have no bill at all.

Central to the debate is what is most cost effective for the owner in terms of total project cost.

Some contracts provide for payments to contractors where errors and omissions in the bill of quantities exceed a certain value. This avoids any difficulties regarding the guaranteeing of bills of quantities. There is an incentive for the owner to be accurate in framing the bill of quantities because of the contractual liability for significant errors or omissions. Other contracts do not work this way.

Arguments advanced by owners who are against having bills include:

- For some projects, bills are unnecessary and in others they are unsuitable.
- The current system of work measurement in use (Standard Methods of Measurement) is defective and needs to be changed.
- Contractors use bills as a basis for extensive and costly claims and disputes throughout a project and following the completion of a project – a part of 'claims engineering'.
- Bills of quantities are not suitable for non-traditional procurement systems where the work is not fully designed and documented prior to calling of tenders.

Arguments advanced by contractors who wish to have bills include:

- Estimating and pricing of work is more efficient with a bill. Without a bill, additional tendering costs are generated.
- Cost control is on a sounder footing.
- Bills form a sound basis for assessing tenders and variations.
- Without a bill, the contractor may have higher exposure; the tender process becomes more expensive.

14.3 TENDER CHECKLIST

As part of the tender examination, a checklist is considered an important management tool for a contractor. The checklist highlights, among other things, matters in the tender documents that need to be qualified in the tender submission. An initial broad brush through the checklist establishes whether or not a tender will be submitted. A finer path through the checklist establishes any qualifications necessary to the tender documents. The following reasons are given to support the use of a checklist:

- It assists in making the tender accurate and complete; the tender, once accepted is the basis for the contract.
- It provides a standardization to the tendering practices of the contractor.
- It transfers an organization's knowledge contained in the checklist to junior staff.
- It is an aide memoire.
- Errors, ambiguities and inconsistencies can be recognized.
- Critical matters can be isolated prior to work starting.
- It assists in pricing the tender.
- It assists in deciding whether to tender or not, or whether the tender needs qualifications.
- It assists in subsequent contract administration.

However, checklists can encourage the users into a 'tick a box' mentality, and have the tendency to give users tunnel vision where they forget about pursuing issues not on the list, and the reasons behind matters on the list. Checklists can reduce individual and lateral thought.

The following, at a minimum, appear on a tender checklist. They are a combination of contractual, management and commercial matters. A generic checklist includes examining:

- General conditions of contract – standard, modified, bespoke; head contract and subcontract compatibility.
- Special conditions of contract.
- Drawings, specification, schedules, program.
- Critical contract conditions – acceptable, needs modification, unacceptable.
- Scope (of work) – location, duration, start date, existing commitments, limits and responsibilities, separable parts.
- Time available for tendering.
- Nominated/not nominated subcontractors.
- Owner competency, past practices and financial standing.
- Payment provisions – types, progress payments, claiming.
- Retention and its release – retained money, bank guarantee.
- Cost adjustment formula – rise and fall.
- Equipment to be provided.
- Approvals – council, authorities.
- Completion and practical completion – dates.
- Extension of time – claim entitlement, claim format, claim timing.
- Delay cost recovery – prolongation, entitlement.
- Dispute resolution clause – appointment of a third party.
- Liquidated damages – entitlement.
- Latent conditions. – entitlement.
- Program – float available.
- Site meetings – attendance.
- Site services – responsibility.
- Insurance.
- Competition – other tenderers.
- Safety obligations.

14.4 RISK SLANT

The above examination and checklist outline many of the issues constituting part of the risk management by the contractor and the owner.

Associated with any input risk source is a magnitude and likelihood, and a downstream output magnitude and likelihood. The output (downside) may be a cost, a delay, damage to the work or third party damage, or injury to a worker or third party. Outputs may be reduced in magnitude and likelihood by prudent action of the owner or contractor.

The origins of the risk sources may be many. They may be a result of actions of the owner, contractor, neither or both. Apportionment of the origins between the parties would be carried out in any attempt at dispute resolution. During the project, however, attempts are made to reduce the magnitudes and likelihoods of the outputs (downsides).

The above sections examine tendering, tender documents and many of the issues which could lead to exposure for the contractor and the owner. Additionally on a project, the contractor has to deal, for example, with the weather, industrial disruption, unclear physical conditions and an unclear market situation. Collectively, these generally are grouped as risk sources, and as having uncertainty.

Analysis and evaluation follow in order to establish the contractor's risk, or owner's risk. A common response of contractors is to lower their risks by subcontracting out responsibilities. A common response by the owner to lower its risks includes thorough checking of the contract documents and incorporating the expertise of designers, constructors and other professionals. Standardization of documentation, with continual improvement over successive projects can assist in risk reduction (Chapter 11).

Conditions of contract produced by standards bodies and neutral organizations and unaltered are generally regarded as having the fairest allocation of responsibilities between the owner and the contractor. This is compared with conditions of contract prepared by industry groups for their members, bespoke conditions of contract prepared by legal representatives for owners, and standard conditions of contract altered by adding special conditions or editing; all can lead to conditions of contract onerous on the contractor. More responsibilities taken by one party could be anticipated to lead to higher risk for that party. There is also a belief that onerous conditions of contract increase the project cost; however, this is difficult to quantify. Interestingly, it has been observed that while contractors dislike having to accept onerous owner contracts, they commonly impose onerous subcontracts on subcontractors. That is, subcontractors are being asked to accept an unfair allocation of responsibilities, yet still perhaps price the work as if the allocation was fair. Project disputation, particularly about variations and being asked unreasonably to take responsibilities, is believed to result from onerous contracts (Chapter 11).

Some further example risk sources that are considered by contractors follow.

Weather. Abnormal weather patterns can seriously disrupt the progress of a project. For example, it is always possible to have a rainfall event that exceeds any reasonable design return period.

Industrial disruption. In order to minimize any industrial disruption, both the owner and contractor acknowledge industrial matters.

Tender conditions may require that the tenderer pay attention to special requirements in the areas of: industrial practices; site agreements; health and safety practices; and site amenities. Local site disputes may be able to be dealt with, while national issues may not be. The size of union coverage, the type of work, for example, building or civil engineering, and the staging of the work are considerations.

Market conditions. There appears to be a direct relationship between the construction market, workload, labour unavailability, decrease in performance and number of contractual claims.

Subcontractors and suppliers. Subcontract and supply prices may make up the bulk of a contractor's bid. These prices have a high potential for variability.

The use of network type representations, such as a fault tree, an event tree or a decision tree, may be of assistance in thinking through the analysis (Chapter 20). They are useful for organizing thoughts, however the actual numbers that arise from analyses using such approaches should not be used unquestioningly.

14.5 CLOSURE

Risk management is particular to each stakeholder at a point in time and situation, and therefore general statements on risk related to any matter cannot be made unless the value system and situation used are common to many people/organizations. Different people/organizations characterize (including, in some cases, ranking and prioritizing) risk differently. Each tender and contract needs its own investigation in terms of risk, and general statements on risk that can be made are few in number, in spite of industry wanting otherwise.

Chapter 15

Project-related risk

15.1 INTRODUCTION

Project-related risk generally relates to some downside exposure on a project. Typically, this exposure would be in terms of:

- Budget overrun (money).
- Schedule overrun (time).
- Work not to specification (quality).
- Accident or death (safety).

Upsides such as completing under budget or finishing ahead of schedule might be included, but commonly these are not explored.

Textbooks speak of one of the tasks for a project manager as risk management. This applies through all stages of a project, including design, implementation and wind down. Risk management is ongoing and the circumstances are continually changing throughout a project's life. Each stage of a project has its own risk sources and these need individual attention; following their identification, analysis, evaluation and response occur related to those particular risk sources. The term control may be favoured on projects instead of the term response. The risk management involves contributions from all affected project parties including design engineers, construction supervisors, operators and contractors. Established project-oriented analysis techniques (or expert judgement in lieu of or in addition to any mathematical techniques) are commonly involved, and these lead to an understanding of the likely exposures during construction.

Projects tend to be one-off and hence present challenges because of this, but practices from previous projects can be used as guidance. Larger projects present greater interest in risk than smaller familiar projects. Investment size, new technology, changing economic conditions and competition increase the interest in looking at risk.

DOI: 10.1201/9781003343592-17

Generally, the importance of risk awareness would be regarded as being greater for the more innovative or unusual projects and greater for larger projects. But whatever the type of project, risk awareness is important. This may mean, for example, for a small project of a similar nature to other projects, that risk management is only undertaken in a qualitative and intuitive fashion. For large capital projects, risk management would generally be undertaken in a quantitative fashion.

Typical projects for which more than a passing recognition of risk should be incorporated include the above-mentioned innovative and capital-intensive projects, new technology projects, fast track projects, projects involving unusual contractual and delivery issues and projects that involve significant environmental, regulatory, safety, political and financial matters.

15.2 WHEN SHOULD RISKS BE EXAMINED?

Risk management is ongoing from project beginning to project completion. That is, risk management is carried out in all stages of a project's lifetime. But, as with most matters pertaining to projects, the work carried out in the early stages of a project could be anticipated to have the most significant impact on the project performing satisfactorily; the influence of decisions could be anticipated to diminish as the project progresses (though this cannot be demonstrated rigorously).

Nevertheless, there are times in a project where identification of risk sources, analysis, evaluation and responding tend to be concentrated. These include:

- The feasibility stage.
- When approvals by the owner/client are sought.
- At tendering.
- During execution/construction/fabrication/…

At the feasibility stage, a knowledge of risks influences the choice of which alternative projects and end-products to proceed with. The feasibility stage is one of the early project stages and has a considerable bearing on the overall cost and direction of a project.

Owner approvals are based on an acknowledgement of risk and the ability to introduce responses. Information on any uncertainty influences the owner's decisions. The owner is provided with estimates for cost and time in the form of ranges or values and associated probabilities.

At the tendering time, a contractor looks at its risk before tendering and committing to a contract with the owner (Chapter 14).

During execution/construction/fabrication/…, new situations may develop which require a re-examination of risk, and perhaps new controls (responses) introduced.

15.3 CONTINGENCY

In estimating and planning, contingencies are commonly used. Other names for contingency are 'cost reserve' and 'time reserve' when talking of cost and duration, respectively. They are sometimes said to cater for 'KUKs' (known unknowns).

Contingencies are applied in the following way. A deterministic cost or duration for a project is calculated. A contingency (an amount of money or a duration) is then added to take care of any unknowns, such as industrial relations holdups or material supply issues. Typically, this contingency is a percentage, for example 10%, but it can be calculated in other ways. An example is a contingency for wet weather. An additional duration may be allowed in the project program and a sum of money relating to that duration may be included in the cost estimate. The duration contingency may be calculated from historical weather data for the particular area, and the costs may be worked out from this duration contingency.

Contingencies are a deterministic idea and hence bear no relationship to risk, though they are usually purveyed in the literature as being a form of risk management. They also cannot be rationally justified, but rather are used because they are quick to incorporate and convenient. They can be thought of as a sort of response in risk management terms. It is commonplace for a fixed percentage contingency (for example, 10%) to be applied to projects without any real justification other than this is what was used on the last project or this is a common industry practice. This percentage might be adjusted up or down depending on the believed uncertainty in the cost or duration estimates.

Contractors, when assembling a bid, might choose to add no contingency if it is believed that by adding the contingency it will make the contractor's bid uncompetitive.

Ideally, any allowance for uncertainty should come from looking at the uncertainty of each project component and assembling this to give an uncertainty for the overall project, rather than something applied in a rigid way solely at the project level. Although still not as rational as something in terms of risk, it is better to see itemized contingencies for each project component because it indicates substantially more thought has been involved.

Contingencies within estimates can be abused. They can be seen as a pot to dip into for various project ills.

15.4 TECHNIQUES AND METHODS

Outline. Typically, projects focus on schedule and cost. Many analysis techniques can be adapted to apply to projects, but the preferred techniques, with respect to schedule and resource (cost) performance, are network based (Carmichael, 2006):

- What-if, sensitivity analysis.
- Monte Carlo simulation.

These methods are illustrated below on an example.

From a planning viewpoint, resources (equipment, people) and method are the fundamental controls (responses) from which cost follows, rather than the commonly believed other way around (Carmichael, 2006). Resources and method are manipulated by the practician, not cost; cost follows from knowing resources and method. Cost is not the fundamental entity even though it might pre-occupy the thoughts of everyone on a project.

Consider the elementary project as represented in the planning network of Figure 15.1 with a single resource. (Complicated projects follow the same thinking as presented here for this elementary project.) The duration (days) and resource (people) requirements for each activity are shown adjacent to the activities. The critical path runs through Admin.-Supply-Execute A-Handover, and the shortest project duration is 21 days.

The bar chart for this project follows from a critical path analysis, and is given in Figure 15.2.

From this bar chart, the resource histogram and cumulative resource diagram result straightforwardly and are given in Figure 15.3. Similar diagrams can be drawn for money – resources get converted to money.

What-if, sensitivity analysis. A 'what-if' analysis looks at schedule and resource (cost) performance (output) when planning assumptions related to resources and method (input) are changed. For example, what if there are delays in design completion, or what if there is an increase in the price of materials?

A sensitivity analysis is more particular, in that it looks at the influence on schedule and resource (cost) performance (output) of small changes in planning assumptions related to resources and method (inputs), rather than new

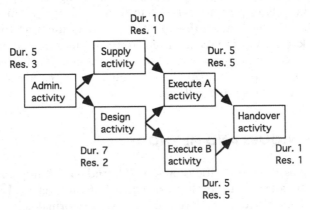

Figure 15.1 Example planning network – activity on node.

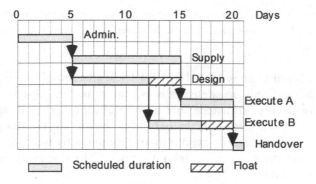

Figure 15.2 Bar chart for example project.

things happening. The matter being studied is said to be sensitive if a small change in an input produces a large change in the output, and insensitive if a small change in an input only produces a small change in the output (Carmichael, 2013a).

Both what-if and sensitivity analyses tend to only change one input at a time, and establish what the change in project output is for this one change. Changing multiple inputs at the same time confuses cause and effect.

A what-if analysis fits within the risk management steps as follows.

- A base case critical path analysis is carried out assuming usual practice. This gives a 'business-as-usual' project completion time and cost.
- *Risk Source Identification.* Risk sources are identified that can lead to the project taking extra time or costing more beyond the 'business-as-usual' case. Examples are delays in material deliveries, shortage of resources (people, equipment), and weather disruption. An estimate of the likelihood or probability of occurrence of each risk source is made. The risk source is interpreted in terms of resources and method.
- *Analysis.* A critical path analysis is carried out incorporating a changed assumption relating to a risk source. For example, if there is a shortage of people to do design work, the duration to undertake the design work will no longer be 7 days, but may be 10 days, 12 days or whatever follows from the availability of resourcing. This gives a new project completion time, and a new project completion cost. The likelihood of the output is estimated from the likelihood of the input carried through the critical path analysis; Monte Carlo simulation might be preferred for doing this.
- *Evaluation.* The risk (corresponding to each risk source) is considered based on the severity of the output (new project completion time, new project completion cost) and the likelihood of this occurring. This evaluation is according to the value system stated in the *Definition and Context* step.

- *Response*. Some action may or may not be taken to address what has just been evaluated. The actions will be in terms of resources and method.

For the project of Figures 15.1, 15.2 and 15.3, an example what-if analysis might be: What if only one designer is available, and the design takes twice as long as originally estimated. That is, the duration for the design activity is 14 days. An examination of the bar chart shows that the design activity had 3 days float, and so the execution and design activities will be delayed and the project duration will extend 4 days to 25 days. Probabilities need superimposing on this.

For the project of Figures 15.1, 15.2 and 15.3, example sensitivity analyses might be: Let the estimates for an activity's resources vary by ±1 person; let the duration of an activity vary by ±1 day, and so on. This will give different project completion times and different project completion costs. The risk management steps are carried out similarly to the 'what-if' case above.

For changes in work order (input), the network is redrawn and new outputs calculated.

Altering activity durations by plus (or minus) some percentage (or resources altered by plus or minus some percentage) leads to altered project durations (or resource quantities). The likelihood associated with output changes is estimated based on the likelihood of the input changes. To obtain probability distributions of the outputs from probability distributions of the inputs, Monte Carlo simulation is a preferred method. Monte Carlo simulation samples from the probability distributions for the activity durations (or activity resources), the network is analyzed, the project completion time and project resources are obtained, the distributions are sampled again, the network is analyzed again, the project completion time is obtained and so on. The method repeats. Refer below and Carmichael (2013a).

Monte Carlo simulation. Monte Carlo simulation applied to networks proceeds by replacing deterministic estimates of activity durations and activity resources (costs) with probabilistic estimates. For each activity, some

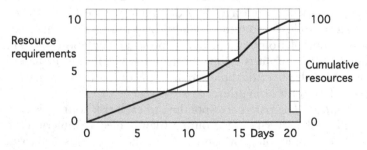

Figure 15.3 Resource histogram and cumulative resource diagram for example project.

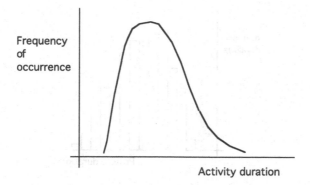

Figure 15.4 Example estimated frequency distribution for an activity duration.

form of frequency distribution estimate is required. Figure 15.4 shows an example for an activity duration.

The source of such frequency distributions can be an issue. Fortunately, the results of Monte Carlo simulation are not overly sensitive to the exact shape of distributions assumed. In many cases, triangular or rectangular distributions can be used without overly affecting the simulation results. The popular ways to establish a distribution are via:

- Historical data, though there may be a shortage of past data.
- Judgement, based on experience and education.
- PERT (program evaluation and review technique) style optimistic, pessimistic and most likely estimates.
- Estimates of means and variances (or standard deviations) and fitting a two-parameter distribution to these.

Monte Carlo simulation then proceeds as follows. This is with respect to schedule. Related calculations can be done on resources (costs) (Al-Sadek and Carmichael, 1992; Carmichael, 2006).

- Each activity distribution is sampled, based on a prior generation of random numbers. This gives a single duration for each activity.
- A critical path analysis is carried out on the network. This produces information such as a project duration, or which activities are critical/noncritical.
- Go to the first Step. The first and second steps are repeated several thousand times, each time producing project information such as project durations or which activities are critical/noncritical. From this, a histogram can be constructed, for example Figure 15.5 for project duration. Or the criticality of activities (the proportion of time over the repeated cycles that an activity is critical) or other information can be calculated.

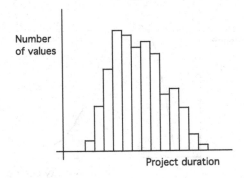

Figure 15.5 Example histogram of project duration.

How such histograms are interpreted in order to complete the *Response* step of risk management is up to the individual or organization, and would be spelled out in the *Definition and Context* step. Values corresponding to ±1 standard deviation from the mean, ±2 standard deviations from the mean, some exceedance probability or other measure might be of interest (Chapter 16).

In effect, Monte Carlo simulation converts one difficult probabilistic analysis into many simple deterministic analyses.

A number of commercially available computer packages are available that do Monte Carlo simulation on project networks. Generally, such packages are add-ons to existing packages that do (deterministic) critical path analysis.

Example. For the project of Figures 15.1, 15.2 and 15.3, a Monte Carlo simulation (with respect to schedule) would proceed as follows:

From estimated frequency distributions for activity durations, each distribution is sampled. Assume the sampling gave the following activity durations:

Admin.	4 days
Supply	9 days
Design	8 days
Execute A	5 days
Execute B	6 days
Handover	1 day

A critical path analysis of the network is carried out. This gives, among other things, a project duration of 19 days.

The above is repeated many times, giving, among other things, many values for the project duration. These values are then plotted in a histogram.

Monte Carlo simulation involves working with probabilities, as does risk. However, many project professionals do not feel comfortable with probabilities and non-deterministic analyses, possibly because of their educational background not exposing them to such thinking.

15.5 APPRAISAL AND FEASIBILITY

Typically, projects (and their end-products) are evaluated for feasibility, or selection between competing projects occurs, based on accepted measures that include (Carmichael, 2014):

- Present worth (net present value).
- Annual worth.
- Future worth.
- Benefit:cost ratio.
- Payback period.
- Internal rate of return.

All these measures are dependent on assumptions (inputs) about:

- Interest/discount rate.
- Benefits and costs (or cash inflows and cash outflows) over time.
- Lifespan of the investment (resulting from the project).

The analysis, from a risk management perspective, uses second-order moment analysis, what-if, sensitivity and Monte Carlo simulation methods, which acknowledge uncertainty in these assumptions. See for example, Carmichael (2011), Carmichael and Bustamante (2014) and Carmichael and Handford (2015).

The analysis input relates to uncertainty in these assumptions, or differing values for these assumptions (inputs) and the likelihood of these different assumptions occurring. The analysis output provides uncertainty information for present worth, annual worth, future worth, benefit:cost ratio, payback period and internal rate of return. Based on the magnitude and likelihood of the output, so the risk level is established. Responses follow.

Chapter 16

P50-P90 project cost estimates - rationale

16.1 INTRODUCTION

Project cost estimates inherently contain uncertainty or variability. The word 'estimate' itself implies uncertainty, and so an estimate is not completely specified by a single number. A distribution of possible costs (outputs) is required to provide a more realistic statement of a cost estimate. The sources of the uncertainty are many, originate pre-project and during projects, and for example include materials supply, resource productivity, resource availability, finance matters, weather, disruptive events, variability in the construction process and prices, varied site conditions, project complexities, lack of information on the future and insufficient historical data. Any decision making about cost estimates needs to acknowledge this uncertainty.

An increasingly common practice at different project stages (phases) is to perform a project cost analysis that reflects the uncertainty in the component project activity costs. The probability distributions assumed for activity costs (estimates) may be based on knowledge of the specific activities as well as experience and past data. Monte Carlo simulation is a favoured analysis tool. An end-product of this analysis is a histogram or frequency distribution for the project cost estimate. Based on the Central Limit Theorem, for many additive activity costs, under reasonably general conditions, this histogram or frequency distribution will have characteristics similar to a normal or Gaussian probability distribution (Benjamin and Cornell, 1970, p. 251; Ang and Tang, 1975) (Figure 16.1). For only a small number of activity costs, the project cost distribution will differ a normal distribution.

A subsequent decision by project personnel involves selecting single deterministic project cost estimates ('point estimates') for budget and communication purposes. Single project cost estimates corresponding to, for example, P50, P70, P75, P80 and P90 costs might be chosen. Here P refers to probability, and Px, x = 50, 70, 75, 80 and 90 among others, refers to the cost which has a probability of x% of not being exceeded (a non-exceedance probability). In some cases, the formalism of the notation Px is not used, but

DOI: 10.1201/9781003343592-18

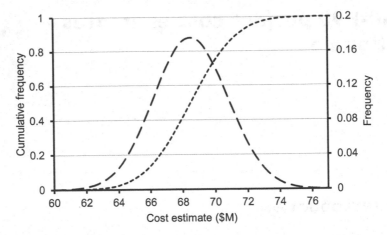

Figure 16.1 Example distribution of project cost estimates; frequency (dashed) and cumulative frequency (S curve) (dotted) diagrams.

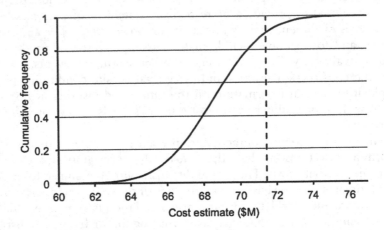

Figure 16.2 P90 value on an example cumulative frequency diagram (S curve).

rather the probability of exceeding x or not exceeding x is quoted. Figure 16.2 shows the P90 value on an example cumulative frequency diagram.

Project cost and funding decisions are generally not concerned with outputs in the extreme tails of the project cost distribution, but rather focus their attention between the 10th and 90th percentiles. The choice of underlying activity cost distribution should not make a large difference to the project costs within this range.

The uncertainty in project cost estimates makes cost-based decisions, particularly funding or budget decisions, difficult. Whether working with the full project cost range or project cost point estimates, uncertainty should be acknowledged. Commonly, decisions are based on information obtained from S curves (cumulative cost curves), with project funding or budgets chosen to match P50, P70 or P80 levels. However, there is no argument that demonstrates that one Px level is more appropriate than another Px level for decision-making purposes. Commonly, the literature talks of these Px values representing different 'confidence levels' interpreted in a lay sense. Some writers believe that budgets should be based on at least a P50 level, but more commonly P70 or P80 levels may be used.

This chapter examines two issues related to this: (i) existing practice connected with Px; and (ii) the basis of the choice of Px, where x = 50, 70, 75, 80 and 90 among others.

This chapter commences with a look at the background to the topic. This leads to an ordered examination of risk, minimum cost, maximum utility and maximum expected utility, all the time examining the rationale for choosing Px.

There are many publications dealing with the analysis of project networks probabilistically, through the use, for example, of Monte Carlo simulation, PERT (Program Evaluation and Review Technique), second-order moment methods and fuzzy sets, including correlation between activity costs. These analyses give information on the probabilistic characteristics of the project cost estimate (Al-Sadek and Carmichael, 1992; Carmichael, 2006). This chapter assumes such analyses as prior knowledge and so this area of knowledge is not reviewed here. Disruptive events of large impact and low frequency (for example, a pandemic, an economic recession, natural disaster or political unrest) seem to be rarely incorporated into any analysis. The estimating process itself may also contain variability. This chapter does not deal with deterministic or probabilistic estimating or network analysis, including sensitivity-style analyses, but rather uses their results, with a particular focus on the resultant uncertainty in project cost estimates. It is noted that sensitivity studies, while straightforward to do, are unable to provide the same information as a probabilistic analysis because no likelihoods are attached to the ranges of activity cost values used.

As a project evolves and becomes better defined, project cost estimates take on reduced standard deviations and possibly a shifting, albeit maybe not large, in the mean project cost estimate. However, this chapter's content is not dependent on any particular stage of a project, or on any standard deviation or mean of a project cost estimate.

The chapter will be of interest to project managers and project owners concerned with the establishment of budget and communication project cost estimates. The chapter offers an original analysis and recommendations.

16.2 BACKGROUND

Existing terminology and practice. With respect to Figure 16.2, the distance from the Px value to the upper value of the cost estimate, a distance measured to the right of the Px value, is sometimes referred to as 'risk exposure', but this is a very inappropriate term and better terms could be used. The distance from a 'base estimate' to the Px value may be referred to as a 'contingency', where the base estimate is a loosely obtained anticipated project cost based on deterministic thinking and deterministic activity costs.

Existing practice first establishes a base estimate. The Px value is obtained by adding a selected contingency percentage to the base estimate. Or the reverse happens, namely that a Px is nominated and this then establishes the contingency percentage. Different Px values represent different degrees of 'confidence', and this 'confidence' is subjective to the decision maker. Public sector bodies may select P80 implying an 80 per cent 'confidence' of obtaining successful (costwise) execution and completion.

Existing practice considers the 'risk exposure', 'base estimate' and 'contingency' as deterministic. Their use represents an expedient approach to a complicated probabilistic situation. However, because they are deterministic, in attempting to deal with something which is inherently probabilistic, it is to be anticipated that such an approach will have deficiencies.

The so-called 'risk exposure', as defined above, goes only part way to establishing what the risk might be. A more complete route to risk is obtained if first both project cost (output) magnitude and project cost (output) likelihood are paired. For any given Px value, the risk associated with exceeding this value is a function of both the cost exceedance magnitude and the cost exceedance likelihood. Further is said on this below.

Contingency. Deterministic project cost estimates attempt to account for uncertainty through conservative adjustments based on the experience of the estimator and the nature of the project, but they will never be able to truly represent the real uncertainty. Commonly, a further percentage is added – called a contingency – in order to avoid the time-consuming work of establishing the uncertainty for each project activity. The contingency percentage could be anticipated to decrease as more details emerge of the project, and also may be adjusted to suit the decision maker's risk preference or attitude.

The intent of a contingency is to allow for uncertainty, in order to avoid over-expenditure of budget, not to allow for changes in scope. A contingency is a financial reserve. Some writers view the contingency as an expression of the risk in a project, but this is incorrect. At best it might be thought of as reflecting risk, but it can never be the risk. Some writers use the terms risk, contingency and uncertainty in a loose and, often wrongly, in an interchangeable fashion. The term 'contingent risk' is also used to impress but not inform.

Using the deterministic concept of 'contingency' is akin to using the deterministic term 'safety factor' in a reliability (probability-based failure) analysis (Chapter 19). Just like using a safety factor is not a rational way of dealing with uncertainty (and has largely fallen from use), so too is using a contingency not a rational way of dealing with uncertainty, particularly when used as percentage adjustments to cost estimates, because it is difficult to assess and account for the multiple presences of uncertainty, with any interaction, that may impact a project and lump these into a single number.

Many publications combine notions of a contingency with Px thinking; however, the two are not compatible – the former is deterministic-based, the latter is probabilistic-based, and they derive from different roots. The relationship between the deterministic notion of contingency and the probabilistic Px value breaks down because the variance or standard deviation of the project cost estimate distribution is itself a variable. A project cost estimate distribution with a low variance will present a quite different percentage contingency to that for a cost estimate distribution with a large variance – the base estimate stays the same but the Px value changes. A similar issue occurs with different assumptions on the underlying activity cost correlations. Nevertheless, for example, at project concept stage, guidance contingency values such as 10% to 15% and 25% to 40% are quoted in the literature as being the same as P50 and P90 estimates, respectively, though no supporting argument is given other than experience, and other publications based on different experiences give different percentages. These percentages would appear to also locate the base estimate in the left tail of the project cost estimate distribution.

Some literature indirectly acknowledges the flaw in linking Px and contingencies, where it is stated that a true Px value can only be calculated through a probabilistic treatment. The incompatible use of both contingencies and Px values in the same cost estimating literature possibly reflects the practice of allowing both traditional deterministic estimates (with adjustments) and more recently promoted probabilistic estimates, and attempting to marry the two in as simple a way as possible. It is also acknowledged in some of the literature that adding percentage contingencies does not assist in understanding project risk.

Choice of Px. When starting with a Px value and establishing the contingency from that, the literature does not indicate on what basis Px is chosen. There is no analysis or justification given as to whether more appropriate Px values should be chosen. Writers appear to accept certain Px values without question. The choice of Px is a commercial decision by the project owner, possibly based on experience or industry convention and said to balance 'risk' and market/economic information, but no transparent justification is given. Some commentary suggests that contractors typically bid jobs around the P50 value, while owners might prefer to have less commercial exposure in respect of capital budgets and instead look for a more conservative P90 value. However, no argument is given in support of these choices.

A large budget Px is chosen to prevent a shortage of funds, to prevent possible delays in administering and approving a revised budget, and presumably to avoid any possible embarrassment from going over budget and subsequently having to request additional funds; yet too large a Px value can lead to a surplus of funds. The Px chosen should ideally balance going under budget with going over budget. A P90 value avoids over-expenditure and fund shortage. It is said to align with actual costs (post project); there is a belief by writers that the actual cost of many projects ends up being close to P90. The question that follows is: Assuming that estimating is carried out appropriately, why are projects anticipated to have actual costs just below P90 when the estimates indicate that the average actual cost of all projects will be P50? In this regard, comment is given below on Parkinson's Second Law type effects.

Against this, particularly if it is a public sector project or one that has to be justified to others, a small Px, or a Px value less than P90, would appear more appropriate. In some publications, the Px value might be said to be discretionary, but still no guidance is given on its choice, other than it be selected 'in line with business requirements'.

Management may also choose to budget at a given Px, but set aside funds at Py, y > x, where the difference y − x represents an additional project cost reserve.

Project owners may nominate particular Px values to use for budgeting and communication purposes, but no explanations are given in the owners' procedure manuals. Some example practices follow: '... entities must use a P50 cost estimate at Stage 1 of the Capital Works Approval Process and a P80 cost estimate at Stage 2'; 'For all projects, the department requires cost estimates to include P50 and P90 values'; '... requires that all estimates are expressed at P90 value or equivalent'; 'The preferred communication is to be estimated project costs P90 and P50'; 'For projects with a value of ..., any potential savings are to be identified using a P75 estimate'; 'An approved business case with P50 and P90 estimates is required for all projects that are intended to be included during the funded years'.

The P50 cost estimate might be used in a benefit-cost analysis, with P90 for sensitivity testing.

The literature loosely links non-exceedance probabilities with an owner's risk management practices, but this is typically done qualitatively. Commentary in the literature may be given that project owners are usually risk averse leading to the use of x values at least equal to 50, though no argument is given as to why Px values greater than P50 are regarded as risk averse, P50 as risk neutral or Px values less than P50 as risk seeking. However, assuming risk aversion starts at P50, arguments are not given as to how much higher than 50 the x value should be. Commonly, investors are considered risk averse (Carmichael, 2013a, 2014, 2020a).

The rationale behind choosing a Px value is examined in this chapter. The chapter offers an original analysis and recommendations.

In other endeavours, beyond project costs estimates, the ideas of P50 and P90 are also used. For example, in renewable energy estimates, banks and investment firms working on wind farm projects may require P50 and P90 values of the wind resource at a location to determine the risk associated with a project's ability to service its debt obligations and other operating costs. However, no justification is given as to why any particular Px values are used.

It is possible that thinking about a project P90 value originates from the simulation practice of modelling underlying activity costs based on three values – a low-side value (activity P10 value), a central tendency value (activity mode, mean or median value) and a high-side value (activity P90 value). These three values are then transformed into a probability distribution for each activity cost, and the distributions subsequently sampled leading to a distribution for the project cost estimate. However, against this, users of PERT style thinking tend to adopt P5 and P95 estimates as optimistic and pessimistic values, when evaluating the moments of an activity cost distribution.

16.3 CASE EXAMPLE

The following sections in this chapter use, as a case example, the road infrastructure project described in Tan and Makwasha (2010). Construction occurs over a 70-week (17-month) period. The P50 and P90 values are given as $68.5M and $71.4M, respectively. With P90 being 1.28 standard deviations from the mean, the standard deviation of the cost estimate is $2.266M, while the variance is this value squared. The probability density function and cumulative distribution function for a normal distribution with a mean of $68.5M and a standard deviation of $2.266M are shown in Figure 16.1. Plus and minus 3 standard deviations from the mean (equivalent to 99.7% of the area under a normal distribution) are $75.298M and $61.702M, respectively. The base estimate obtained by non-probabilistic means is $47.7M to which a percentage contingency is added giving a cruder and different P90.

The estimated cost versus x, based on a standard normal probability distribution table, is shown in Figure 16.3. The plot is nonlinear because of the

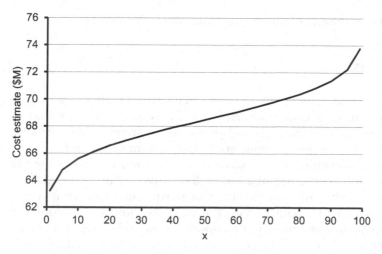

Figure 16.3 Estimated cost versus x for the case example.

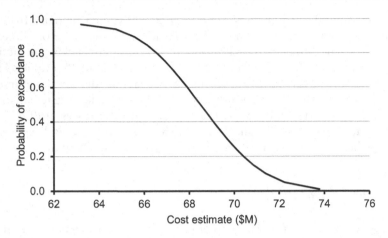

Figure 16.4 Probability of exceedance versus cost estimate.

tails of the normal distribution. Figure 16.4 shows the probability of exceedance versus cost estimate.

Using this case example, the following sections look at what is the appropriate choice of Px value based on underlying reasons different to those discussed above, namely:

- Risk.
- Overspend.
- Loss of access to funds.
- Total cost – utility.
- Contingency – utility.

A table at the end of the chapter summarizes the choices.

16.4 RISK

The literature loosely links non-exceedance probabilities, Px, with an owner's risk management practices, but this is typically done qualitatively and without transparency. Risk only exists in the presence of uncertainty. Risk is related to the pair {output magnitude; output likelihood}, where output refers to something about exceeding cost. Each owner will view risk associated with cost differently, and each project may be viewed differently in terms of risk. Risk is something that is particular to each owner in each different set of circumstances. Primarily the downside – cost overrun – occupies most attention, but an upside – cost saving – is also possible.

For any given Px value, the risk associated with exceeding this value is related to the pair {cost exceedance magnitude; cost exceedance likelihood}.

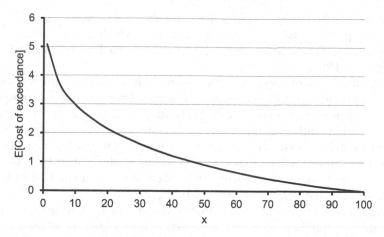

Figure 16.5 Expected cost of exceedance versus x for the case example.

Here, magnitude is taken as the mean of the area of the frequency distribution to the right of the Px value, and likelihood is taken as the area under the frequency distribution to the right of the Px value. These two values when multiplied give the expected cost of exceedance, denoted E[Cost of exceedance], where E[ ] refers to expected value. This is plotted in Figure 16.5 for the case example.

Commentary in the literature may be given that project owners are usually risk averse. Risk-averse owners might be anticipated to only use Px values greater than P50. Figure 16.5 suggests that to be truly risk averse, owners should choose a Px value as large as possible, but the plot does not indicate any particular Px value as being the most appropriate.

16.5 OVERSPEND

As noted above, a large budget Px may be chosen to prevent a fund shortage (from over-expenditure) and to prevent possible delays in administering and approving any revised budget. P90 might be used for this purpose. There is a belief that P90 aligns with anticipated actual cost. However, there is also the view that the actual cost of a project ends up being close to whatever Px value is used, whether this is P90, P80 or other.

There is much literature on the cost performance of projects. While it can be difficult to relate budget with actual cost because of matters such as design change variations and assumptions changing, there is a view that many projects cost more than originally budgeted even though a direct one-to-one comparison between the budgeted amount and actual amount may not be possible. Some cost overruns can be attributed to estimating practices.

Setting any Px value, and particularly for Px values greater than P50, as a budgeted project cost introduces the situation which can be described in terms similar to the 'self-fulfilling prophecy' or Parkinson's Second Law (Parkinson, 1960).

The self-fulfilling prophecy, in the socio-psychological literature, refers to a prediction or expectation coming true simply because people believe that it will and people's resulting behaviours align to fulfil that belief. That is, people's beliefs influence their actions. A belief or desire, correct or incorrect, could bring about a certain outcome. In project terms, setting a high Px value as the budget will thus lead to the project actually costing Px. However, the literature's intent of choosing a high Px value is that the budget is not exceeded, not that the actual cost will be Px.

Parkinson's Second Law states that 'Expenditure rises to meet income'. Like the self-fulfilling prophecy, it could be interpreted as whatever budgeted value of Px is set by the project owner, then the project expenditure will rise to meet this value.

Related thinking, but in a time not cost sense, are Parkinson's First Law (Parkinson, 1958) and the notion of uneconomical drag out on projects (Antill and Woodhead, 1970; Carmichael, 1989a, 2006). Parkinson's First Law states that 'Work expands to fill the time available for completion'. The amount of time that a person has to perform a task is the amount of time that person will take to complete the task. Interpreted in terms of cost, then expenditure expands to meet the budgeted Px amount. Uneconomical drag out is often mentioned with reference to project compression or project crashing. Compressing an activity below some normal (least cost) duration comes with a cost penalty; likewise lengthening the duration of an activity also comes with a cost penalty, sometimes called uneconomical drag out. Using a larger budget Px value may induce uneconomical drag out.

A belief that the actual project cost will in fact approach the nominated Px value is contained in project manuals. Any difference between a Px estimate and the actual cost is regarded as an 'error'. Attaining the Px value as the actual cost is viewed as desirable.

If it is assumed that by setting the budget at Px the project cost will naturally rise to Px in the sense of Parkinson, then something like Figure 16.6 may apply (using P50 as the starting point for over expenditure). The dotted curve is expected cost of exceedance, the dashed line represents the overspend (beyond the P50 cost), here assumed uniformly spent beyond P50, and the solid curve is the sum of these two components. Figure 16.6 suggests that owners should choose P50 as a Px value. Of course, different assumptions on the overspend and when it starts will move the minimum point to other Px values.

Figure 16.6 assumes 100% overspend, that is the actual project cost will reach the Px cost. Reducing the percentage overspend leads to Figure 16.7. The plots in Figure 16.7 range from 10% to 18% overspend, and are given over a lesser range of Px values than Figure 16.6.

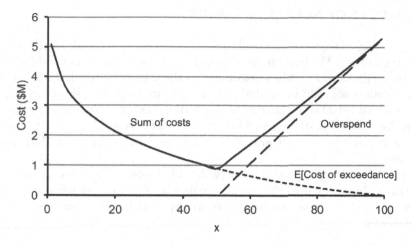

Figure 16.6 Expected cost of exceedance and overspend combined; 100% overspend.

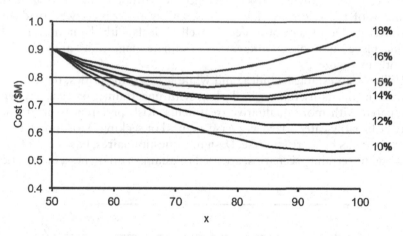

Figure 16.7 Expected cost of exceedance and overspend combined; different % overspends.

The minima in Figure 6 correspond to the following Px values (with percentage overspend in brackets): P70 (18%), P75 (16%), P80 (15%), P85 (14%), P90 (12%) and P95 (10%). And hence, according to any assumption made on overspend, so the appropriate Px value will change. Commonly assumed values of P80 and P90 correspond with percentage overspends (beyond P50) of approximately 15% and 12%, respectively. Of course, different assumptions on overspend and when it starts will move the minimum points to other Px values.

16.6 LOSS OF ACCESS TO FUNDS

Prescribing a larger Px may lead to more funds being set aside by the owner for the project. The extra funds required could be regarded as the difference between P50 and Px. When discounted to the start of the project, their present worth (PW) follows the dashed line in Figure 16.8. Also shown in Figure 16.8 by a dotted line is the present worth of the expected cost of exceedance, and by a solid line the sum of both present worths. With different interest or discount rates, the diagrams do not change much in shape, but do in magnitudes.

Figure 16.8 suggests that owners should choose a Px value as small as possible, but does not indicate any particular Px value as being the most appropriate. If now, overspend is incorporated, this result is further reinforced.

16.7 TOTAL COST – UTILITY

Knowledge of an owner's risk preference or attitude can add to an understanding of the selection of Px values. Utility provides a way of dealing with risk preferences or attitudes. As well, it deals with the nonlinear value people place on money – different cost estimate magnitudes are viewed differently by owners.

Utility is an established idea, and while it has some critics, there is large agreement on its usefulness in dealing with uncertainty, particularly when compared with more qualitative treatments. Risk preferences or attitudes may be broadly categorized as averse, neutral or seeking. Most project owners are seen as being risk averse. Designed questionnaires, based on different cost and uncertainty scenarios, are used to establish an owner's utility curve,

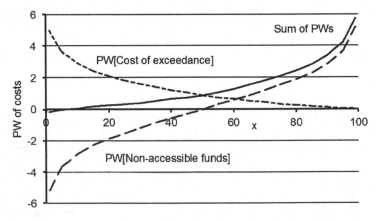

Figure 16.8 Present worth (PW) of non-accessible funds and cost exceedance.

and the utility curve or function shape indicates the owner's risk attitude (Ang and Tang, 1984). For each different project cost situation and uncertainty, while also taking into account the market, any competition and the circumstances of the owner (investor), a new utility curve or function would be derived. A utility curve shows an owner's relationship between utility (vertical axis) and any particular outcome (horizontal axis) – here project cost. Risk averse, neutral and seeking follow concave, straight line and convex shapes, respectively (Ang and Tang, 1984). Figure 16.9 shows typical utility curves for a range of levels of risk aversion, as well as the risk-neutral case (RA = 0). Higher utility is favoured over lower utility. In Figure 16.9, the values for project cost are scaled to give a range from 0 to 1. As mentioned, most people tend to be risk averse, but the below approach applies to any risk preference.

Utility curves may be described mathematically in a number of ways, for example, with an exponential function or a quadratic function, u(C), depending on the user's preference. Here u is utility and C is cost. The establishment of utility curves is well documented in the literature (Ang and Tang, 1984) and is not repeated here. The following figures are based on quadratic utility functions of the form,

$$u(C) = \alpha C^2 + \beta C + \gamma$$

where u is utility, and α, β and γ are constants, different for each investor and investment circumstance. The second derivative of u gives an indication of the risk attitude. The degree of risk aversion, RA, is sometimes measured by,

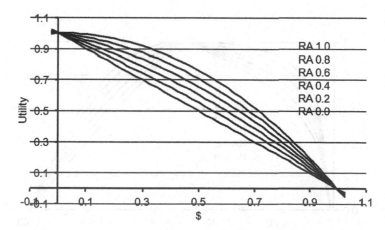

Figure 16.9 Typical utility curves (functions), showing different levels of risk aversion, used in this 'chapter's numerical studies. Risk aversion coefficients shown in the legend. Increasing risk aversiveness from bottom to top.

$$RA(C) = -\frac{u''(C)}{u'(C)}$$

The expected value of utility, using a Taylor series expansion, leads to (Benjamin and Cornell, 1970; Ang and Tang, 1975; Carmichael, 2014, 2020a)

$$E[U] \cong \alpha E^2[C] + \beta E[C] + \gamma + \alpha Var[C]$$

where E[ ] and Var[ ] denote expected value and variance, respectively. Such an expansion is valid for usual utility function shapes. Ang and Tang (1984, p. 74) note that *the expected utility is relatively insensitive to the form of the utility function at a given level of risk-aversion, and that the expected utility does not change significantly over a wide range of risk-aversion coefficients. Hence, the exact form of the utility function may not be a crucial factor in the computation of an expected utility. Moreover, the risk-aversiveness coefficient in the utility function need not be very precise; that is, any error in the specification of the risk-aversiveness coefficient may not result in a significant difference in the calculated expected utility.*

The desirable result for the project owner is one that which maximizes expected utility.

Figure 16.10 shows the expected utility of the total costs shown in Figure 16.6, that is where there is 100% overspend in the style of Parkinson. Var[C] is assumed small. Figures 16.11 and 16.12 show the expected utility of the total costs shown in Figure 16.7 for the overspend case of 10% and no overspend, respectively.

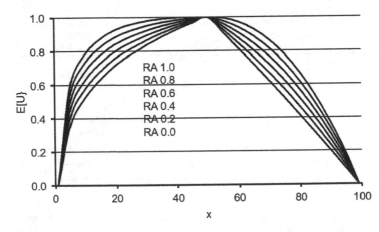

Figure 16.10 Expected utility of the total cost shown in Figure 16.6; 100% overspend.

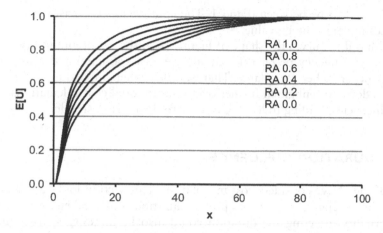

Figure 16.11 Expected utility of the total cost shown in Figure 16.7; 10% overspend.

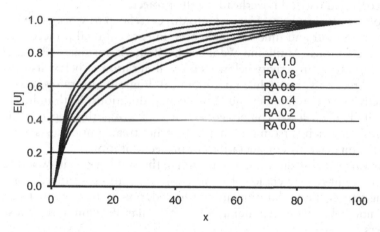

Figure 16.12 Expected utility of the total cost shown in Figure 16.7; no overspend.

For the 100% overspend case, Figure 16.10 suggests a Px value ranging either side of P50. For the 10% overspend case, Figure 16.11 suggests a Px value ranging from below P50 upwards. For the no overspend case, Figure 16.12 suggests a similar Px value to the 10% overspend case. The conclusions are sensitive to the owner's level of risk aversion.

16.8 CONTINGENCY – UTILITY

The value that an owner places on the contingency can be studied through looking at the utility of the contingency. The utility curves of Figure 16.9 can be used where now the horizontal axis is the dollar value of the contingency, normalized to be between 0 and 1. The contingency is measured from the base estimate to the Px value.

With utility curves the shape of Figure 16.9, the maximum utility occurs with a contingency of zero (Px equals the base estimate) and a minimum utility occurs when Px is large. That is, utility when applied to contingency values does not enlighten further over conventional cost calculations. Low contingencies, leading to low Px values are the more desirable.

16.9 DURATION INFLUENCE

All of the above considers cost in isolation. This section looks at the influence on the choice of Px when project duration is also taken into account. Uncertainty affecting the duration could also be anticipated to influence uncertainty in the cost estimate, while duration magnitude and cost magnitude are also linked. For example, extra duration may contribute to escalation costs and may add overheads to the project.

Project costs can be categorized into direct (related to work quantity) and indirect (one-off and time varying) costs. In conventional project compression calculations (Antill and Woodhead, 1970; Carmichael, 1989a, 2006), on compressing an activity below its normal duration or below its least cost value, direct costs rise, while uneconomical drag out also raises direct costs for activities extending beyond their normal durations. With indirect costs typically being a lesser component of the total costs, this suggests that the conclusions reached from this chapter's earlier treatment of costs alone will be little impacted by considerations of project duration.

For various cost-duration relationships, the simulations in Wood (2002) shed no light on the choice of Px as project durations change. This is not saying that cost and duration are not linked, only that looking at duration does not help in understanding why particular Px values are chosen by owners.

This can be confirmed by looking at expected utility. The owner, in principle, wishes to maximize expected utility, which can be approximated by (Benjamin and Cornell 1970; Ang and Tang, 1975; Carmichael, 2014, 2020a)

$$E[U] \cong \alpha E^2[C] + \beta E[C] + \gamma + \alpha \mathrm{Var}[C],$$

where E[ ] and Var[ ] denote expected value and variance, respectively.

For general cost-duration (C-D) relationships, the inclusion of duration, D, offers no change in understanding of what Px value should be used by owners. The choice of Px is not influenced by any considerations of project duration, but rather only by cost magnitude.

16.10 SUMMARY

For budget and communication estimates, deterministic approaches establish a base estimate and add a contingency, while probabilistic approaches go directly to a Px value. There is no relationship between these deterministic and probabilistic approaches although they commonly appear in the same project guide documents, and often as alternatives. The use of contingencies is an expedient attempt at addressing uncertainty in project cost estimates but it has flaws. Contingencies and Px values should not be anticipated to be related because of their different origins and underlying assumptions.

There is no quantitative way of demonstrating what Px value an owner should use. Depending on the underlying reasoning, different Px values result. Any Px value, from small to large, could be justified depending on the project owner's emphasis. There is no single rationale. Table 16.1 demonstrates the dilemma, based on the case example, and associated assumptions used in the chapter.

Table 16.1 Px resulting from different underlying reasoning, based on case example

Underlying Reason	Px suggested
Political – to avoid a request for extra funds	Large Px
Political – to sell a project	Small Px
Risk aversion – minimum cost exceedance	Large Px
Risk aversion and 100% Overspend – Parkinson type effect	P50
100% Overspend – Parkinson type effect	P50
18% Overspend	P70
16% Overspend	P75
15% Overspend	P80
12% Overspend	P90
Loss of access to funds	Small Px
Loss of access to funds and Overspend	Small Px
Total cost – Utility; 100% Overspend – Parkinson type effect	Either side of P50
Total cost – Utility; 10% Overspend	Below P50 and Px > P50
Total cost – Utility; No Overspend	Below P50 and Px > P50
Contingency – Utility	Unchanged from Cost Estimate
Duration influence	Same Px as without duration

Chapter 17

Options

17.1 OVERVIEW

Risk within options is typically discussed in the literature in a qualitative way. An option value, being an expected value, would appear to not involve risk; however, by looking at the components that make up the option value, it can be shown that the risk treatment of options contains the same elements as those outlined in Part A of the book.

Chapter 3 provides that as a precursor to establishing risk, pairs of output information are required – {output magnitude; output likelihood}. Here likelihood may be the probability associated with an output or an output exceedance (or non-exceedance) probability, noting that output magnitude may have a sample space. The full probability distribution for the output is not considered because of the multiple combinations of statistical moments (central tendency, dispersion, skew and kurtosis) which are possible.

Particular applications in dealing with risk might report output in ways other than as the pairs {output magnitude; output likelihood}:

- Either output magnitude or output likelihood is reported for a constant value of the other.
- Expected value of the output, for so-called 'risk-neutral' attitudes.
- Expected utility of the output, where the utility measure is that of the stakeholder for whom the risk study is being carried out – different stakeholders will have different utility functions.

17.2 OPTION VALUE AND RISK

An option value is calculated as a single expected value (of output) and not in terms of separate magnitudes and likelihood (of the output). In particular, an estimate of an option value is given by (Carmichael et al., 2011; Carmichael, 2014, 2016b, 2020a)

$$OV = \Phi M$$

DOI: 10.1201/9781003343592-19

Here, $\Phi = P[PW > 0]$ and is termed 'feasibility' (Carmichael and Balatbat, 2008; Carmichael, 2014, 2016b), P[ ] is probability, and M is the mean of the present worth (PW) upside and is measured from the origin (Figure 17.1). This equation has been called the Carmichael equation to distinguish it from the Black-Scholes equation. The larger the value of Φ, the more viable an investment becomes. Where $\Phi > 0.5$, there may be no need for an option calculation because deterministic calculations might already indicate viability (subject to any up-front cost).

For applications, see for example Carmichael et al. (2015, 2019a).

An option value is an expected output value. To understand expected output values in terms of risk requires knowledge of the stakeholder's value system. To work this through it can help to think in terms of a matrix similar to those mentioned in Chapter 7, where the matrix has 'axes' of Φ and M.

> Different combinations of Φ values and M values can lead to the same ΦM value, but do not necessarily lead to the same risk level. The option value, ΦM, is not risk unless the stakeholder chooses it to be that way.
>
> Two stakeholders looking at the same investment may calculate the same option (ΦM) value, but associate different risks to this value based on different value systems and the particular component Φ and M values (which in turn result from the shape of the present worth upside distribution).

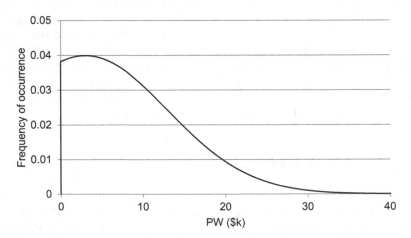

Figure 17.1 Example upside of the present worth (PW) distribution; Φ is the area under the curve; 'mean of PW upside' is the mean of the area under the curve measured from the origin.

Chapter 18

Accounting, finance and economics

18.1 INTRODUCTION

The usage of the term risk in the accounting, finance, economics and related disciplines literature takes on multiple meanings, and can be incompatible with usages in other disciplines as well as within their own. Given the large number of existing publications in accounting, finance, economics and related disciplines, all doing their own thing, it may be too late to correct, but hopefully not.

The requirement to assess risk, disclose risk and conduct risk management is becoming non-negotiable in accounting, finance, economics and related disciplines. Risk impacts public policy debates. Standards and good practice now require deep consideration of risk. However, interpretations of the term risk vary across different applications. All current usages have a similar intent and many refer to future underlying unpredictability and a tendency for actual outputs (consequences, outcomes) to differ from the anticipated outputs; however, there is no consensus as to its precise meaning. This is demonstrated here in a comprehensive review. The book suggests that for the state of the art on risk to develop, for communications and disclosures on risk to be meaningful and in-depth, and for standards to be respected, some agreement will be necessary across applications on a preferred meaning for the term.

18.2 BACKGROUND

Risk has become a ubiquitously used term in accounting, finance, economics and related disciplines, and impacts public policy debates. Risk has become ever-present in professional practice, as it has in everyday life. It is a part of all corporate reporting and decision making, whether overtly or not. Consideration of risk and risk management is seen as becoming more important in order to fully understand and direct business operations, it has become key vocabulary used in organizations, and

DOI: 10.1201/9781003343592-20

it influences debates and organizational and management practices for both the public and private sectors.

Increasing rates of change in technology, markets, public awareness and climate for example are necessitating this increased emphasis on risk. Standards, discipline guidelines and discipline discussion papers, stock exchanges and regulators are placing increasing emphasis on risk, risk reporting and disclosures, verification and risk management. This is occurring in not only conventional accounting, finance and economics practices but also in emerging thinking on sustainability, climate change and other non-financial areas; these emerging areas impact financial decisions and create the need for greater transparency.

Risk type or categories affecting accounting, finance, economics and related disciplines may range broadly, for example: insurance, credit, liquidity, market, financial, interest rate, currency, price, inflation, actuarial, investment, market, natural, strategic, commercial and operational, along with for example, in emerging areas, physical, environmental, social, climate, transition and regulation. However, shortcomings in the way risk is managed within the accounting, finance and economics sectors have been noted in some publications.

The term risk is seen to be interpreted, used and defined in multiple different ways. Risk can have different meanings to different individuals. While all current usages have a similar intent and may refer to future underlying unpredictability and a tendency for actual outputs (consequences, outcomes) to differ from the anticipated outputs, there appears to be no consensus in its precise meaning. There is no widely accepted definition of the term, leading to ambiguity and differing interpretations. The question that arises is: How can communication on risk, which arises within guidelines, policies and practices, be facilitated through removing some present ambiguity, while acknowledging the necessary role of subjective and judgemental elements? This chapter reviews the multiple interpretations and usages of the term and the potential dilemmas, and raises the issue of reconciling all the different meanings in order that communications, disclosures and reports on risk are interpreted in a consistent fashion by all users.

Common meanings found in the literature (accounting, finance, economics and related disciplines) for risk include 'chance', 'possibility', 'event', 'impact', 'likelihood', 'uncertainty', 'probability', 'standard deviation' and 'variance', but there are many more (listed below and Chapter 3). Preference for usage changes between and across practice areas. Yet generally these are different concepts and conflicting, potentially creating confusion. Use in a technical sense may differ from that given in dictionaries. Besides causing potential confusion and preventing proper communication on risk, this chapter also argues that such meanings do not allow an internally consistent development of risk management. No one appears to have attempted any reconciliation of the different understandings of risk in accounting, finance, economics and related disciplines. The question that arises is: How can risk be reported and understood, and risk management carried out, if everyone has a different meaning for risk?

The state of affairs appears perplexing. It is unclear as to why no one has attempted to correct the current situation, or at least acknowledge the incompatibilities of the various meanings. Interestingly, based on the author's experience: some people are unaware of the multiple meanings used for risk and hence had never considered the need for reconciling different meanings; while some people believe multiple meanings are natural to cover multiple situations and feel no need to develop one tight meaning for risk enabling clarity of communication or to change the status quo. The latter group believe that risk is highly contextual, and people should be able to differentiate different meanings in any situation. Such people appear comfortable communicating with a word that can mean something different to a sender and a receiver. However, it would seem desirable in any profession that everyone agrees on definitions for terms used in order that the profession can advance.

When discussing risk, most attention seems to focus on negative outputs, downsides or losses and their associated likelihoods, possibly because people are typically risk averse and like to avoid downsides. The term *downside risk* might also be used. This gives negative connotations to risk. Unless it is what is referred to as 'pure risk' (downsides only), it need not be this way. It is also often possible that positive outputs, upsides or gains can occur; however, this tends to receive much lesser attention or is treated separately to risk. Likelihoods may possibly be both objective and subjective, and arise from any suitable source. Risk may also be used as a verb or noun, but below reference is mainly to the noun.

The chapter looks at the predominant usages of risk in accounting, finance, economics and related disciplines, through a review of the literature, and highlights the array of meanings that exists. The literature reviewed is very broad and includes not only journals, but also industry documents, standards, regulations and discipline dictionaries. The chapter outlines the plethora of current meanings for risk but also suggests a way forward that fits within a rational approach to risk management. The intent of the chapter is to encourage discussion on a way forward for addressing risk in a consistent way across multiple applications.

While there are many papers incorporating various aspects of risk, few have stopped to first look at the lack of consensus as to risk usage, and the conflicting meanings. Of the papers that have looked broadly at risk and its meaning, these tend to have a reliability, safety and engineering focus (Carmichael, 2016a), rather than an accounting, finance, economics and related disciplines focus. Some publications give a limited historical account, some look at the 'risk-return' relationship, while some look at risk perception. There is a significant gap in knowledge on this matter. Lay usage may favour qualitative meanings, while professional usage may favour quantitative usage. The interpretation of risk appears to be context and background dependent.

The chapter is structured as follows. Firstly, terms and their differences are outlined. The chapter then highlights the various different usages of risk in accounting, finance, economics and related disciplines, with some commentary. The usage of risk in company risk registers and the output magnitude-output likelihood relationship is then explained. The literature touching on risk is immense. As such, the chapter extracts example uses of risk to highlight the state of the literature and practices. The chapter does not look at risk analysis, assessment, culture, or appetite; the chapter solely deals with risk interpretation, usage, and meaning which feeds into these risk activities along with risk management.

The chapter gives an original treatment of the multiple interpretations and usages of the term risk and the potential dilemmas, and raises the need to reconcile different meanings in order that communications and reporting on risk are consistently interpreted by users. The chapter offers an original review and analysis; there appears to be nothing similar existing in the literature. The chapter will be of interest to practitioners and researchers in accounting, finance, economics and related disciplines.

18.3 TERMS

The terms given here are those used in this book's argument on risk and risk management. See also Chapter 1.

In this book, 'possibility' is used in the sense of a sample space (with probability attached), that is a value in a collection of many conceivable values. The values could relate to inputs, outputs or other depending on the application. 'Probability' and 'chance' are used in the same sense to indicate relative 'likelihoods', frequencies or weightings attached to these values. Probabilities, chances and likelihoods may derive from some objective root, for example data, or be subjectively established and can arise from any suitable origin. The mention of likelihoods acknowledges 'uncertainty' (implying variability or variation), as opposed to determinism. Risk only exists in non-deterministic environments; something is regarded as 'risk-free' when a certain (equivalently, no uncertainty) output exists. 'Variance' is 'standard deviation' squared and is regarded as a more fundamental quantity than standard deviation, though both terms convey similar information; both are measures of dispersion or deviations from the mean. The unit of measurement for standard deviation is preferred by many over that for variance. Range also implies dispersion. Interestingly, the use of the non-dimensional coefficient of variation does not seem to occur in the risk literature. 'Expected value' is equivalent to mean or average; all are measures of central tendency. Variance, standard deviation, coefficient of variation, expected value, average and mean are precisely defined within treatises on probability and statistics.

In discussing risk, there is an originating input (source), which leads to an output (consequence, outcome). The upstream input results in a downstream output. For example, a source may be a currency fluctuation and this comes with likelihoods or uncertainty; the downstream output of this may be a financial loss with the attached likelihood or uncertainty usually assumed to carry through from the source/event, but this carry-through need not apply. Other examples are: the value of bonds (the output) depends on interest rates, reinvestment, inflation, default, ratings and liquidity (sources), all of which contain uncertainty; a general investment's output may be expressed as profit, return on investment, present worth, internal rate of return or payback period or other similar measures, and these depend on interest rates and cash flows (sources) which contain uncertainty. The interchangeable use of thinking on inputs (sources) and outputs (consequences, outcomes) is commonly seen, highlighting that many people do not understand their fundamental differences. Many originating inputs can be present, each with their own likelihoods. Different inputs could be regarded as possibilities (within a collection of inputs), each leading to different outputs, which themselves could also be regarded as possibilities (but now within a collection of outputs).

18.4 USAGES IN ACCOUNTING, FINANCE, ECONOMICS AND RELATED DISCIPLINES

18.4.1 Outline

The literature in accounting, finance, economics and related disciplines is reviewed here and the various risk usages, definitions and conflicts noted. Where explicit definitions are not given in the literature, the meaning can be interpreted from the context. A comparative analysis is given of the different usages and meanings, with similarities and dissimilarities noted.

It is seen that writers have used many terms – including possibility, probability, chance, likelihood, uncertainty, variance, standard deviation and expected value – to describe risk. These usages are also reflected in dictionaries. As such, their continued usage is natural, but as outlined below, their usage is not helpful in communications. The book argues that all these terms have a similar intent and many imply future underlying unpredictability and a tendency for actual outputs to differ from anticipated outputs, but that they offer only part of the picture for risk management purposes, and that a more complete picture is required.

A lot of the literature in accounting, finance, economics and related disciplines could be tidied up and simplified if the definition of risk put forward in this book was adopted. A lot of the complexities presented in the technical literature disappear, and the literature would appear less esoteric with the use of this book's definition of risk. However, it is understood that many academics have made a career and name out of such complexities, and so there will be resistance to change.

The following sections are presented in terms of the co-existing ways that risk is interpreted, used or defined, namely:

- Possibility, loss, impact.
- Chance, probability, likelihood, uncertainty.
- Standard deviation, variance, volatility, range.
- Variations, fluctuations, changes, deviations.
- Expected value, utility.
- Exposure.
- Risk as risk.
- Miscellaneous.
- Output magnitude and likelihood.

Commonly, different silos of knowledge give different definitions. All the usages of risk outlined in this chapter have a common purpose, and that is to aid reporting and decision making primarily in monetary transactions, but risk associated with non-monetary transactions is possible. The intent of the definitions, it could be argued, is the same but strictly their meanings are different and incompatible. Present in many usages is a reference to the future and underlying unpredictable tendency for actual outputs to differ from the anticipated outputs. There is sparse commentary in the literature regarding the conflict arising from these multiple meanings.

It is seen that the literature tends not to favour any one particular interpretation, but rather there is a large range of preferred usages. Some publications adopt more than one meaning and hence are not internally consistent. All the described usages offer a part picture of risk and associated risk management, but not the full picture which is discussed later in the chapter, and which also offers the opportunity for internal consistency. Risk could be thought to embody most of the above bullet point meanings, but generally not any one individual meaning.

It is also seen that layperson perceptions of risk intrude on the usage and understanding of risk in professional disciplines, and professionals may adopt lay usages for risk when more precise usages would be more helpful. Various common sayings used in both a lay sense and professional sense include: *at risk, at your own risk, run/take a risk, bear a risk, put at risk, risky, riskier, riskiness, risk shifting, risk factor* and *spread your risk*, and all can imply different meanings. Adages and idioms on risk abound, as discussed in Chapter 3. And all give different interpretations of risk. The literature mentioning risk is vast.

All the above bullet point meanings present confused thinking on risk. This confusion is outlined here. It should be contrasted with the book's view presented in Part A, so that the current confused thinking in accounting, finance, economics and related disciplines can be highlighted.

18.4.2 Possibility, loss, impact

Where the term *possibility* is used in the literature, it can refer to the upstream input or the downstream output. In the latter case, the output may be measured relative to some anticipated value. At times, the word *possibility* is implied. Chance is described in Part A as being equivalent to probability. However, in lay terms, it may be used in an equivalent sense to possibility. In this last sense, it is similar to *potential*. Possibility and probability are seen to be used interchangeably in parts of the literature in line with their lay meanings, though their mathematical meanings are very different. The terms *impact*, *result* and *loss* may also be used, all with negative connotations.

18.4.3 Chance, probability, likelihood, uncertainty

Probabilities and likelihoods can come from available (objective) data or subjective weighting. Probability implies uncertainty, though lay (wrong) usage may be different. Chance in Part A is described as being equivalent to probability; however, lay usage may (wrongly) differ.

The expressions *risk-reward, risk versus return, risk-return, risk-return relationship, risk and return* or *risk/return tradeoff* are commonly used without rigour. A cornerstone in finance theory *is the relationship between risk and return* (in spite of this strictly being circular); *people are said to only accept higher risk if they get a higher return.* Here risk is used to mean a qualified probability; higher positive rewards or returns should coincide with, and would be desired, where there are lower probabilities of getting a positive return, and vice versa. (Risk here may also be interpreted as standard deviation or variance as covered below.) The popular interpretation of risk gives the reason for the paradox of Bowman on the risk-return relationship. Greater uncertainty is accepted in return for a potentially greater return.

With the expression *risks and opportunities*, opportunity is considered positive and risk negative. A reward, return or opportunity is usually viewed in a positive sense, but a more general interpretation around risk is that both positive and negative outputs are possible, while the inclusion of the word reward, return or opportunity is not required if the meaning for risk of Part A is adopted.

The expressions *risks, opportunities and outcomes, risks and uncertainties*, and *risk, opportunities and financial impact* are also seen.

The popularly written about *risk-free rate* coincides with outputs being certain, that is any probabilities are ones and a deterministic situation holds. Risk cannot exist without uncertainty. The *risk-adjusted rate* is based on the risk-free rate and a *risk premium*. For the risk-averse person, the risk premium is adjusted upward (downward) if the uncertainty is perceived to be high (low). *The riskier the investment, the higher the return the investor needs.* Adjusting the risk premium upward (downward) has the effect of reducing (increasing) the present worth of any future cash flows. Such an adjustment to the risk-free rate is a pragmatic approach at simplifying life and providing something that is simple to understand, but strictly is not rational and its naming is misleading; a risk-adjusted rate cannot simultaneously deal with uncertainty and the time value of money (Carmichael, 2017).

18.4.4 Standard deviation, variance, volatility, range

Standard deviation squared is variance. Both standard deviation and variance measure dispersion, as does range. Volatility may be derived from standard deviation, or it may also be used in a less precise sense to mean a tendency to change rapidly and unpredictably. Much of the finance literature adopts standard deviation, variance and volatility as measures of risk. Reducing standard deviation, variance or volatility is considered desirable in many situations, it being viewed as reducing the level of uncertainty.

18.4.5 Variations, fluctuations, changes, deviations

The terms *variation* and *fluctuation* may also be used by some writers to describe risk, both referring to the presence of different values, but not in as a precise sense such as variance or standard deviation.

18.4.6 Expected value, utility

The term 'expected' might be used in a dictionary sense to mean anticipated, or in the precise mathematical sense of expected value, corresponding with average or mean. The latter is obtained from the product of output magnitudes and output likelihoods.

A *risk-reward ratio* might be calculated in terms of expected loss divided by an expected return. If now, an investor's utility function (indicating risk aversion, risk neutrality or risk seeking) is incorporated, then the ratio is one of expected utilities. Risk registers and output magnitude-output likelihood thinking, mentioned later in this chapter, can be and are used to obtained expected values.

18.4.7 Exposure

Risk is sometimes spoken of as an *exposure*, without precisely defining what that means, but generally referring to some downside.

Occasionally the combined words *risk exposure* is seen, as well as *exposure to risk*.

18.4.8 Risk as risk

When risk is defined in terms of itself in a circular way in the literature, its meaning may range over all the other given alternatives.

18.4.9 Miscellaneous

There are many other less common usages of the term risk, and these muddy the waters further. There is also *embedded risk*, for example, *risk of a risk*.

18.4.10 Output magnitude-output likelihood

Output magnitude and output likelihood, as a precursor to understanding risk, are also mentioned in Sections 18.3 and 18.5. Nothing is said in the literature about the stakeholder's value system and the connection of the pairs {output magnitude; output likelihood} to risk, as defined in this book.

18.4.11 Discussion

A range of conflicting usages for risk can be seen in the literature, including conflict within documents. There is seen to be no universal agreement on the meaning of risk in accounting, finance, economics and related disciplines. All usages have an underlying intent commonly relating to losses and anything that may lead to loss, but all have different meanings. Present in many usages is a reference to the future and unpredictability, a tendency that actual output may differ from the predicted output. Although maybe implied, in many cases, there is no direct reference to output likelihoods or to any range of output magnitudes (losses, gains).

The meaning given to risk by writers determines what can be done under any subsequent risk management. Some meanings paint only a part picture of risk and hence only permit a restricted approach to risk management, and therefore such meanings can be argued as being deficient.

Risk and risk management are intertwined, but risk management only proceeds after firstly knowing the risk. The meaning of risk as a function of both magnitude and likelihood of any output would appear to allow most scope for full risk management. Such a meaning incorporates most thinking on risk as it relates to usages such as possibility, probability, standard deviation and variance as outlined. Risk management then reduces to one of influencing the inputs (magnitude and likelihood) or the relationship between the inputs and outputs, and hence influencing the resultant risk in order to achieve desired or acceptable risks according to an objective function(s).

There appears to be something not overt happening with people's understanding of risk at the same time as the term is being used. Users seem to be ignoring the strict meaning how risk is being used, and perhaps adopting a more qualitative understanding in given scenarios. However, the question remains: How it is possible that professionals communicate when their meanings for risk are different?

Despite the range of meanings, accounting, finance, economics and related disciplines continue to function, and people continue to communicate. The literature is quite mature now, and perhaps it is too late to make changes, and too late to get people to change their habits. Humans appear to have an innate resistance to change, and there is the saying 'if it isn't broken, don't fix it', but clearly the terminology usage is seriously broken. Reconciliation of all meanings may now be difficult, given the distinct differences across and within applications, but not without hope.

18.5 RISK REGISTERS AND OUTPUT MAGNITUDE-OUTPUT LIKELIHOOD

Risk registers appear commonly within organizations in order to address risks associated with internal organizational practices and external influences on the organization, including matters that impact company cash flows and profits. In conjunction with output magnitude-output likelihood charts (also presented as a matrix) they provide a better basis (but not the whole picture) for a rational treatment of risk and risk management, in comparison to many other risk-related treatments. Risk registers and output magnitude-output likelihood charts are not totally without criticism but their underlying intent is sound.

Risk registers take the form of tables where the table columns variously comprise: a description of the risk type; input (source of the risk); likelihood of occurrence; output (consequence); risk level; available responses or controls to manage the risk; and sometimes an allocated responsibility within the organization. Such registers are a useful and readily understood communication tool within and beyond an organization showing how risks are being addressed, and they are updated over time. Each input (risk source) is listed separately, but there may be interdependencies with other listings. Entered risk levels derive from output magnitude-output likelihood charts, and these levels are anticipated to reduce as responses or controls are introduced. Both positive and negative outputs could be noted in the register, but more usually only negative outputs are focused on. The risk register makes explicit the difference between the input (source) and the associated output (consequence) but is silent as to how an output derives from its input, and possibly the output likelihood is assumed to be the same as the input likelihood.

An output magnitude-output likelihood chart or matrix displays risk levels within the chart or matrix against 'axes' of output magnitude and output likelihood. The chart or matrix is silent, unfortunately, as to how the likelihood of an output derives from the likelihood of its originating source. The scales chosen for the 'axes' are at the discretion of the user, and may be quantitative, qualitative or a combination of the two, and span any anticipated ranges of output magnitudes and output likelihoods. The approach accommodates both qualitative (subjective) and quantitative views of risk. Output likelihood is given on a frequency scale, while output magnitude is given on a severity scale. The scales may change with different risk types, while qualitative scales introduce subjectivity because of the imprecision of language. Generally, the larger the output magnitude and the higher the output likelihood, the higher the risk level, and vice versa, but nothing is said about why this is so or about the stakeholder's value system. The charts are a useful and readily understood communication tool within an organization showing how risks are perceived by the organization.

Both risk registers and output magnitude-output likelihood charts adopt the definition of risk as a function of the pair {output magnitude; output likelihood}; the output derives from an input with attached likelihood. Risk as a function of the pair, {output magnitude; output likelihood}, embodies all the usages seen in the literature and referred to in Section 18.4. Both elements of the pair are necessary. The level of risk (alternatively, this might be thought of as level of management responsibility), converts the pair based on the value system of the relevant stakeholder (typically an organization), including their risk attitude or *risk appetite* and level of risk aversion. That is, the same two numbers for {output magnitude; output likelihood} can represent different risk levels to different stakeholders. Different stakeholders are seen to have different perceptions of risk, and hence the notion of a transfer of risk unchanged between stakeholders is not possible. The measurement scale particulars for risk are not fixed, but rather differ from stakeholder to stakeholder, situation to situation and risk type to risk type. Risk can only exist in the presence of uncertainty (here reflected in likelihoods).

The definition of risk as a function of output allows common thinking about all the possible ways that risks come about and any associated risk categorization, which can be added later by users if they wish. This risk definition allows any risk type (credit risk, financial risk, climate risk, ...– as mentioned above in Section 18.1) to be dealt with, without modification; for example, it is not necessary to monetize the output; both monetized and non-monetized risk situations, and both qualitative (subjective) and quantitative approaches, are handled in the same way.

18.6 RISK MANAGEMENT

By adopting risk as a function of {output magnitude; output likelihood}, the risk management steps can be developed comprehensively with internal consistency, and across all risk types or categories. Typically, risk management is described to go through the steps of: *Definition and Context*; *Risk Source Identification* (magnitudes and likelihoods); *Analysis and*

Evaluation; and *Response*, and is done iteratively with feedback to earlier steps in an attempt to establish the best response. This requires knowledge of a stakeholder's value system leading to a statement of an objective function(s) and constraints. So-called dynamic risk management sees this practice repeated through time. So-called enterprise risk management sees the practice broadened to organization level. Management becomes risk management.

The *Definition and Context* step establishes the situation and value system of the relevant stakeholder; this leads to the measurement scale particulars (qualitative or quantitative) for risk based on output magnitude and output likelihood pairs, the objective function(s) by which a best response (adjustment, control) is selected later in the risk management steps, and any constraints on this selection. Different attitudes to risk (different degrees of risk appetite or of risk aversion, neutral or seeking favoured by different stakeholders) lead to different measurement scale particulars for risk. That is, risk management involves subjectivity and risk and risk management are peculiar to each stakeholder. The commonly thought of notion of risk perception is embodied directly. A commonly used objective function for selection of a response is one of 'minimizing risk' when risk is perceived as a downside only, but other objective functions are possible. The identification step relates to risk sources, though it is frequently (but wrongly) referred to as identifying risks. Analysis (*Analysis and Evaluation* step) converts the inputs (magnitude and likelihood) into outputs (magnitude and likelihood) and this might be done qualitatively or quantitatively. Commonly, a one-to-one conversion of input likelihood to output likelihood is assumed in order to simplify thinking, but it need not be assumed that way. Evaluation (*Analysis and Evaluation* step) interprets these outputs in terms of the previously established scales for risk. Lastly, the *Response* step attempts to target the inputs (magnitude and likelihood) or transformations between the inputs and outputs in order to address the previously established objective function(s) and constraints. Responses may lead to risk elimination or reduction, leaving the risk unchanged, or other.

18.7 CONCLUSION

The chapter highlights the large discrepancies in interpretations of risk across accounting, finance, economics and related disciplines. This impacts public policy debates and communication. The intent of the different understandings of risk, it could be argued, is the same but strictly their meanings are different and incompatible. Present in many understandings is a reference to the future and an underlying unpredictable tendency for actual outputs to differ from the anticipated outputs. Using risk as a function of the pair, {output magnitude; output likelihood} embodies most of the meanings reviewed, but not necessarily any one individual meaning. The large range of risk meanings may have the effect of retarding the development of and practices in risk thinking in these disciplines, and hence policy debate.

There does not appear to be any study made in the accounting, finance, economics and related disciplines literature on the many interpretations of the term risk, and consequently no reconciliation has been attempted. A review of the literature highlights the difficulty that could be faced in attempting to create a single acceptable meaning across all usages. Given the extensive development in the literature in areas that incorporate the term risk, and long histories of adoption of using the term risk, the reconciliation of the meanings of risk may represent too great a challenge. Ideally, a single meaning for risk would be desirable. Communication on risk would benefit. The theory in these disciplines would simplify. It would seem desirable in any profession that everyone agrees on definitions for terms used in order that the profession can advance.

Some people are not aware of risk being given multiple meanings, and when made aware, do not believe that the different meanings have impaired their understanding in communications, or believe that reconciliation is necessary. Others argue that risk is contextual and different meanings are to be anticipated in different situations, and people adjust in order to communicate and understand; accordingly, having multiple meanings is not a great concern to these people. Nevertheless, this requires an educated user if switching contexts occurs. People who work in silos are also not affected by usages outside those silos. The counter-argument is that professional disciplines only develop on the back of accepted single meanings for core terms, and not on the back of lay, multiple interpretations of terms. Communication could always be anticipated to be an issue when the sender and the receiver have different interpretations of words.

The impetus for the reconciliation of terminology, if considered desirable by the profession, would need to come from professional industry bodies and consensus of their members. Reconciliation in a technical sense is possible, but this first requires the acceptance of the need for reconciliation or any elimination of resistance to reconciliation by professionals, and this may be harder to achieve.

This chapter suggests a possible way forward using a definition of risk as a function of {output magnitude; output likelihood}. This also permits internal consistency within risk management and a rational structured treatment of risk management. Some current usages of the term risk referred to above satisfy parts of this book's view, but not always all of it. The definition allows any risk type or category (credit risk, financial risk, climate risk, ...– as mentioned above in Section 18.1) to be dealt with without modification; for example, both monetized and non-monetized risk situations, and both qualitative (subjective) and quantitative approaches, can be handled in the same way.

Chapter 19

Reliability

19.1 INTRODUCTION

Reliability ideas are typically developed in the literature in terms of 'the reliability of systems', and in terms of 'state' and 'capacity'. Systems are made up of subsystems, elements or components. Reliability results accordingly have wide applicability. 'State' is a measure of system behaviour, while 'capacity' is a measure of the system's resistance, with both expressed in the same units of measurement.

Reliability is the probability that a system will perform (that is, will not fail). Let p_s be the probability that the system fails, then $(1-p_s)$ is the reliability of the system.

Failure probability together with failure magnitude (for example, as a cost) lead to risk in line with the stakeholder's value system.

Prior to the trend towards reliability thinking, safety factors were used to acknowledge uncertainty. However, being a deterministic approach, safety factors could not be used to establish risk.

19.2 BACKGROUND

Reliability, in a general sense, refers to a system's ability to perform a specified task for a specified time in a specified environment. It is measured as a probability.

One of the roles of technologists, engineers, ... is to assure performance of the systems under their responsibility. Nevertheless, systems do fail to perform as intended, with possible extreme manifestations of breakdown and collapse. It is argued that because of incomplete information and the presence of uncertainty, there will always be a finite probability of non-performance of any system. It may be too costly to make this probability negligibly small.

Consider a building structure as an example. The structure's safety or reliability depends on the anticipated loadings (resulting in the state) over the life of the structure and the materials, member sizes, ... (collectively giving capacity) used. The loading cannot be predicted with certainty. Even the material characteristics, member sizes, ... contain variability. The absolute assurance of safety of the structure is therefore not possible.

It is interesting to note that because structural engineers have done their job so well over the years, structural failures are rare, and the community (and some engineers) have developed a false belief about the permanence of society's structures.

For a water supply system, the available supply is variable and dependent largely on the weather, while the usage or demand fluctuates. The absolute assurance of adequate water to consumers is therefore not possible.

In terms of design, reliability may be set as a constraint, for example, a system with a reliability exceeding a specified level; or as an objective function, for example, a system with maximum reliability. Codes of practice may be written in terms of systems achieving a certain level of reliability, though on the surface, explicit mention of reliability may not be made.

Availability and reliability are linked. Provided the system is functioning, then it is available for use. User requirements may be in terms of availability, which in turn reflects on reliability. Reliability is an important part of operation and maintenance in possibly most circumstances.

For establishing the reliability of complicated systems, it may be necessary to develop it from component/subsystem reliabilities.

This chapter is structured in terms of static systems and dynamic systems. Within static systems, the following issues are discussed: reliability measure, reliability block diagrams, series systems, parallel systems and combined series-parallel systems. Within dynamic systems, the following issues are discussed: extended view of reliability, reliability over time, continuous time, failure rates, series systems and parallel systems.

There are numerous publications on reliability. The body of knowledge has been developed to a reasonable degree of sophistication. Treatments from both a heavy mathematical and pragmatic background can be found. Reliability is a field where mathematics, and in particular probability and statistics, has made considerable in-roads into practice, though in the implemented form the mathematics may be heavily disguised in order to make it acceptable to people from all backgrounds.

19.3 STATIC SYSTEMS

A static system is one whose behaviour is independent of time.

Reliability measure. Reliability is the probability that the system will perform. This requires the introduction of a measure of system behaviour. Here the term state is used to describe behaviour (but depending on the situation, the term demand might be used). The symbol X is used to denote the state. Also required is a measure of system capacity. (This might be referred to as resistance or supply by different writers depending on the situation.) The symbol Y is used to denote the capacity. Typically, the capacity of a system may be established by prior testing to failure and failure analysis of like systems.

For a system to perform, the capacity, Y, has to exceed the state, X. Both X and Y are measured in the same units. Both state and capacity are random variables and so reliability is defined as the probability that the capacity exceeds the state, that is

$$\text{Reliability}, R = P[Y > X]$$

Failure occurs when the capacity is less than the state, that is

$$\text{Probability of failure}, p_F = P[Y < X] = 1 - R$$

Former approaches to assessing system performance centred on the notion of a 'factor of safety'. This looked at preset, usually worst case, values for both the state and capacity and ignored variability. More recent thinking is in terms of a reliability measure which acknowledges full variability of both state and capacity.

To calculate p_F, it is necessary to look at the overlap between the density or mass functions (probability distributions) for Y and X. The overlap depends on the shapes of the density or mass functions and their relative positions, that is, essentially the means and variances of the two distributions. p_F can be evaluated through integration involving the two probability distributions for Y and X.

From where do the probabilities and probability distributions come? Typically, they come from historical data, test data, opinion or estimates. In many cases, complete probabilistic information may not be available on the state or capacity, in which case an expert's 'guesstimate' may be included, or the analysis might be reduced to a second order analysis – only dealing with expected values (means, averages, central tendencies) and variances (standard deviations, coefficients of variation, dispersion) (Carmichael, 2014, 2020a).

Margin of safety and safety factor. A safety margin, M, may be defined as

$$M = Y - X$$

A safety factor may be defined as the ratio Y/X. These typically use average values or deterministic estimates for Y and X. They are deterministic measures compared with reliability which is probabilistic.

Multiple failure modes. Strictly the above treatment of reliability applies to single failure modes. Multiple failure modes in a system may also occur. This is distinguished from a system of multiple components each of which may fail. A building structural member, for example, may fail in multiple failure modes – shear, flexure or by buckling.

The establishment of possible failure modes in a system may be carried out with the help of fault trees or event trees (Chapter 20). Generally, safety of a system implies that none of the failure modes occurs; failure implies one or more of the failure modes occur. Where failure modes are correlated, some form of approximation is necessary in the failure analysis.

Redundancy. Redundancy may be included in a system in order to improve its reliability. Reliability is dependent on the degree and type of redundancy. In systems without redundancy, failure of a component implies system failure.

Redundancy might be categorized as:

- Active.
- Standby.

In an active-type redundancy, all components are sharing the load. In a standby-type redundancy, some components do not come into play until failure occurs elsewhere in the system.

Reliability block diagrams. Systems are made up of subsystems, elements or components. The reliability of the total system might be established from the reliability of its subsystems. Reliability block diagrams (RBDs) assist in this regard. The diagrams reflect the reliability aspects of the system, and not necessarily the physical structure of the system. Each block represents a subsystem or component. The block is like a switch – either closed (functioning and current flowing) or open (failed, and current not flowing). Examples of reliability block diagrams are given below.

Series systems. Where failure of one component means failure of the whole, it can be represented as a series system. Series systems have no redundancy. When the weakest link in a chain fails, the whole chain fails. Schematically series systems may be represented by Figure 19.1.

Let,

E$_i$ the failure event of component i
E$_s$ the failure event of the system
p$_i$ the probability that component i fails
p$_s$ the probability that the system fails
R reliability

Then $(1-p_s)$ is the reliability of the system, R$_s$, that is the probability that the system will perform or function properly. Non-failure of the system follows from non-failure of all the components, that is,

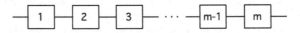

Figure 19.1 Series system representation.

$$\bar{E}_s = \bar{E}_1 \cap \bar{E}_2 \cap \dots \cap \bar{E}_m$$

Or for statistically independent events,

$$R_s = \prod_{i=1}^{m} R_i$$

$$R_s = (1 - p_s) = \prod_{i=1}^{m} (1 - p_i)$$

This is sometimes referred to as the product rule of reliability – the reliability of the system is the product of the reliability of the individual components. It implies that the reliability of the whole is less than the reliability of each component (because each R_i is less than 1). The system reliability decreases rapidly as the number of series components increases (Figure 19.2).

The approximation,

$$\prod_{i=1}^{m} (1 - p_i) \cong 1 - \sum_{i=1}^{m} p_i \quad \text{for} \quad \sum_{i=1}^{m} p_i \ll 1$$

is good for high R_i and small m.

Parallel systems. Where failure of a component in a system does not necessarily mean failure of the whole system, this might be represented by a parallel system. A parallel system contains redundancies. Figure 19.3 shows schematically a parallel system.

In contrast to series systems where survival is given by the intersection of component survival, failure in parallel systems is given by the intersection of

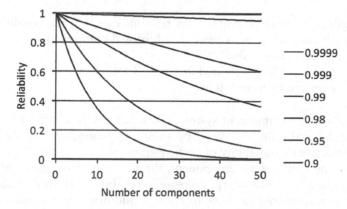

Figure 19.2 Influence of increasing number of series components/subsystems. Reliabilities of each component are noted in the figure.

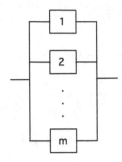

Figure 19.3 Parallel system representation.

the component failures. Failure of the system follows from failure of all the components, that is

$$E_s = E_1 \cap E_2 \cap ... \cap E_m$$

Or for statistically independent events,

$$p_s = \prod_{i=1}^{m} p_i$$

This is sometimes called the product rule of unreliabilities. It is a similar result to that for series systems but with 'probability of no failure' replaced with 'probability of failure'. Unreliability decreases (reliability increases) as the number of parallel components increases.

The system reliability becomes,

$$R_s = 1 - p_s = 1 - \left(\prod_{i=1}^{m} p_i \right)$$

Example. For a component with a reliability of 0.8, the reliability of a system made of such components, as additional components are added in parallel, increases as in Table 19.1.

The system reliability is plotted in Figure 19.4.

Some observations from this are:

- For a many-component system, the reliability is a series of 9s (after the decimal point) followed by some other number. It is difficult to appreciate the reliability of such systems.
- Starting with a single component, the addition of an extra component in parallel provides a large increase in reliability. For further additions, there is a diminishing benefit. There is additional cost of adding additional components, but they may not return an equivalent benefit in terms of reliability.

Table 19.1 Parallel system example

Number of components	System reliability	Incremental reliability	Percentage comparative reliability
1	0.800000	—	—
2	0.960000	0.160000	20.00
3	0.992000	0.032000	24.00
4	0.998400	0.006400	24.80
5	0.999680	0.001280	24.96
6	0.999936	0.000256	24.99

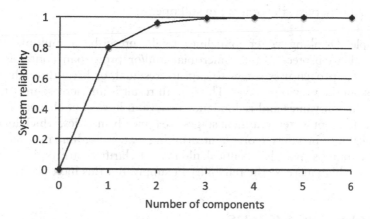

Figure 19.4 Parallel system – change in reliability with number of subsystems/ components.

Combined series-parallel systems. Commonly systems have characteristics which are neither purely series nor purely parallel systems, but rather components with mixed series and parallel connections. To analyse such systems, they are broken or aggregated into subsystems of series components, and subsystems of parallel components, and then combined at the subsystem level until a single subsystem exists.

Examples. Consider some common systems:

- A suburban water distribution network (with respect to water supply). A suburban water distribution network provides water to end users. A combined series-parallel system might be used. The main line uses a parallel or network system in case of failure. The sub-lines use a series system to distribute water from the main line to every home. This gives simple pipe installation, low-cost investment and modest maintenance.
- A suburban electricity distribution network (with respect to electricity supply). A suburban electricity distribution network has a similar

combined series-parallel system as a water network. The main line uses a parallel system and the sub-lines use series systems.

- A computer network (with respect to connecting users). Unlike the previous networks (water, electricity) which have one-way flow from a source to the users, the computer network needs two-way communication between a source (server, provider) and users or between users. A parallel system connects the user to a server or a network hub (sub-network). So, when there is a failure in one user, the whole system still keeps working. For some large networks, the interconnection between hubs and servers (main network) uses a series system to provide a larger network with simple installation.
- A lift/elevator. Building lifts are a parallel system in that if one lift is out of service, the others can still operate.

Example. Drinking water guidelines might prescribe a multiple barrier approach to protect against microbial and/or protozoan contamination. Such an approach uses stages to capture residual, or breakthrough pathogens from the previous stages. The overall result is an increase in pathogen log-reduction values and redundancy provision in case of element failures. Hence different water treatment stages are typically in series, which increases the reliability in safety (compliant drinking water). Redundancy hardware such as pumps, parallel identical filters and clarifiers are typical parallel systems which increase the reliability in supply. But this increases the cost.

19.4 DYNAMIC SYSTEMS

A dynamic system is one whose behaviour is time-dependent.

Extended view of reliability. Reliability, in a general sense, refers to a system's ability to perform a specified task for a specified time in a specified environment. It is measured as a probability.

A system functions until it fails, though the exact point as which failure is said to occur may not be able to be defined precisely. For something that breaks suddenly, it is easy to define the point of failure. For other systems, they may gradually deteriorate. Failure occurs, then, in two possible ways:

- Through drift, for example, wear and tear, ageing or fatigue.
- Suddenly, for example, breakage or catastrophic failure.

However, it is convenient to think of a system having only two possible states:

- Functioning satisfactorily.
- Failed.

For more involved treatments of reliability, further states can be defined related to the period where the performance is drifting.

The age at which a system fails may be expressed, quite naturally, in terms of a time unit, for example, hours, days, years, ..., but may also be expressed in terms of some other evolving unit, such as distance travelled or cycles completed.

The environment includes everything except the system (Carmichael, 2013a). Specific environments may be more relevant in defined reliability calculations, for example, temperature, humidity, dust, wind, chemical environment or operator characteristics. That is, only part of the total environment is looked at, the rest not being considered relevant.

Reliability over time. Table 19.2 illustrates some test data on a mechanical component. In total, 100 components were tested, the last failing during the period 1500–1600 hours.

In Table 19.2,

i period number, i = 1, 2, ..., 16;
f_i number of components that failed in period i;
F_i cumulative total of failures by the end of period i;
  number of components that have failed by the end of period i

Table 19.2 Example failure data.

Period (100 hours)	f_i	F_i	R_i	λ_i
1	0	0	100	0.000
2	1	1	99	0.010
3	1	2	98	0.010
4	2	4	96	0.020
5	3	7	93	0.031
6	11	18	82	0.118
7	17	35	65	0.207
8	23	58	42	0.354
9	13	71	29	0.310
10	9	80	20	0.310
11	9	89	11	0.450
12	6	95	5	0.545
13	2	97	3	0.400
14	2	99	1	0.667
15	0	99	1	0.000
16	1	100	0	1.000
	100			

$$F_i = \sum_{j=1}^{i} f_j$$

R_i total number still functioning at the end of period i

$$R_i = 100 - F_i$$

λ_i number of failures in a given period as a proportion of the number still functioning at the beginning of the period.

$$\lambda_i = f_i / R_{i-1}$$

Reliability comes directly from dividing R_i by 100 – the proportion of components still functioning at the end of period i. For example, at the end of period 5, after 500 hours, the reliability is 93%.

The probability of failure by the end of a period i is given by dividing F_i by 100. For example, at the end of period 5, the probability of failure is 7%.

The probability of failing in a given period is given by λ_i. For example, the probability of a component failing in the fifth period is 3.1%, assuming it is still operating at the start of the fifth period.

Figure 19.5 plots f_i, F_i, R_i and λ_i.

Continuous time. Where time is continuous rather than being represented as discrete periods, the relationship between failure frequency, reliability and failure rate is as follows.

Probability of failure. The probability that the time of failure lies between time a and time b,

$$P[a < t < b] = \int_a^b f(t)dt$$

Failure function,

$$F(t) = \int_0^t f(t)dt$$

Reliability,

$$R(t) = \int_t^\infty f(t)dt \quad \text{or} \quad f(t) = \frac{-dR(t)}{dt}$$

$$= 1 - F(t)$$

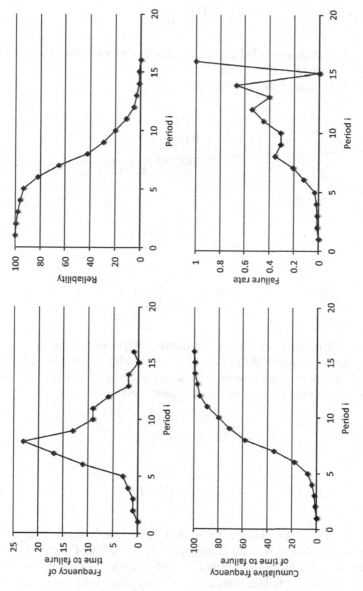

Figure 19.5 Plots for the example.

R(t) is the probability that the system is still functioning by time t, or that it fails after time t. R(t) may be called a survivor function.

Failure rate,

$$\lambda(t) = \frac{f(t)}{R(t)}$$

If the system has not failed by time t, then the probability that it will fail in the interval t to t + dt is $\lambda(t)dt$.

Combining some of these expressions

$$\lambda(t) = \frac{f(t)}{R(t)}$$

$$= \frac{-dR(t)}{dt} / R(t)$$

or

$$\frac{1}{R(t)} dR(t) = -\lambda(t)dt$$

$$R(t) = \exp - \int_0^t \lambda(t)dt$$

That is, if $\lambda(t)$ is known, then the reliability R(t) can be found.

Mean time to failure (MTTF). In place of reliability functions, it may be more convenient to work in terms of a single measure, such as mean time to failure – the average lifetime of many systems, given by

$$MTTF = \int_0^\infty tf(t)dt$$

$$= \int_0^\infty R(t)dt$$

Many systems have failure rates that follow a bathtub shape curve. Plotting failure rate versus time, the bathtub curve had three phases: early life failures – high failure rate initially with decreasing failure rate; regular operation – constant failure rate; and wear out phase with increasing failure rate. There are several types of bathtub curves (RBTC – reliability bathtub curves) which may be constructed.

The reliability curves for different products will be different – for example, mass-produced speakers and subsystems for an aircraft – the gradient on the first and third phases of the curves, and the length of the middle or the useful life sections will be substantially different.

The bathtub curve is a ubiquitous characteristic of living creatures as well as of inanimate engineering devices, and much of failure rate terminology comes from studies of human mortality distributions.

Systems are composed of multiple subsystems, sub-subsystems and so on. For the system to function, it requires all lower-level components to function. For the initial part of the bathtub curve (infant mortality), the system components are being tested for the first time. The middle part of the bathtub curve corresponds to usual system operation and has the smallest and nearly constant failure rate. It describes the 'useful life' of a system with random component failures. The last part of the bathtub curve corresponds to ageing and wear and tear taking its toll as the system nears the end of its life expectancy.

Knowing the bathtub curve shape, maintenance can be scheduled accordingly over the life of a system, and manufacturers can offer warranty periods accordingly.

Example. Most construction projects have durations up to a few years. Operation over a product lifecycle is not so critical in these short-duration activities because equipment is not anticipated to operate continuously for periods comparable to its operating life. The issue is that of the failure rates of equipment as the equipment progresses through the first section of the curve, prior to entering the constant failure rate region. It is common for major equipment to be re-built between projects because the contractor cannot afford to have downtime due to failures when in the execution stage of a project. Once re-built, the contractor has to 'burn-in' equipment to ensure that it operates as intended when required. The shape of the curve is altered by things such as over-stressing. This can be the case with the contractor's equipment. It is worked hard when it is being used, and hence the requirement for re-builds. For this region, the contractor attempts to progress the equipment through the burn-in phase and into the constant failure rate portion of the curve by running in the equipment prior to deploying.

Failure rates. To make the mathematical treatment easier, it is convenient to assume some standard mathematical expressions for the failure rate.

Constant failure rate (CFR)

$$\lambda(t) = \text{constant} = \lambda$$

This leads to

$$R(t) = \exp(-\lambda t)$$

$$f(t) = \lambda \exp(-\lambda t)$$

$$F(t) = 1 - \exp(-\lambda t)$$

$$MTTF = 1 / \lambda$$

The advantages of using a constant failure rate are:

- Mathematical ease.
- λ is estimated straightforwardly from 1/MTTF.
- Fits many systems during their early operation/maintenance period.

The disadvantages of using a constant failure rate are:

- May be hard to fit data with only one parameter (λ) to vary.
- Assumes no ageing effect, wear and tear effects, etc.

Plots of these expressions are given in Figure 19.6. (The plots correspond to the $\beta = 1$ case.)

Failure rate varying with time; the Weibull distribution. A particular case of where the failure rate is assumed to vary with time is obtained if the Weibull distribution is assumed for the failure density function

$$f(t) = \frac{\beta}{\alpha} \left(\frac{t}{\gamma} \right)^{\beta-1} \exp- \left(\frac{t}{\gamma} \right)^{\beta}$$

Here β is referred to as a shape parameter and γ as a scale parameter (or characteristic life). The basis for selecting values for β and γ is discussed below.

This Weibull distribution assumption leads to:

$$\lambda(t) = \frac{\beta}{\gamma} \left(\frac{t}{\gamma} \right)^{\beta-1}$$

$$R(t) = \exp- \left(\frac{t}{\gamma} \right)^{\beta}$$

That is, the failure rate λ varies as a power of t.
The advantages of using a Weibull distribution are:

- Easier to fit data, with two parameters (compared with CFR with one parameter).
- Weibull distribution based on a 'weakest link' notion – the system fails when the weakest link fails. Many systems behave similarly.

$\beta = 1$ corresponds to an exponential distribution.

$$f(t) = \frac{1}{\gamma} \exp- \frac{t}{\gamma}$$

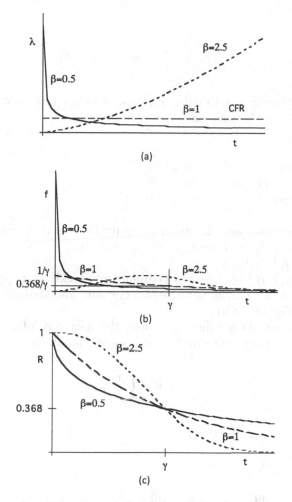

Figure 19.6 Based on Weibull distribution: (a) Failure rate; (b) Lifetime distribution; (c) Reliability.

and

$$\lambda(t) = \frac{1}{\gamma} \quad (\text{constant})$$

$\beta = 2$ corresponds to a Rayleigh distribution.

$$f(t) = \frac{2t}{\gamma^2} \exp\left(-\frac{t^2}{\gamma^2}\right)$$

and

$$\lambda(t) = \frac{2t}{\gamma^2} \quad \text{(linear in time)}$$

Values of β are chosen to reflect the type of failure rate anticipated. From Figure 19.6, it can be seen that:
When

$\beta < 1$, failure rate is decreasing (DFR)
$\beta = 1$, failure rate is constant (CFR)
$\beta > 1$, failure rate is increasing (IFR)

Plots of these expressions are given in Figure 19.6 for different values of β.
Some further characteristics of these curves are:
When $t = \gamma$, $R = 0.368$
For large β, R is large (close to 1) and drops quickly as it approaches γ.
Failure rates for distributions besides constant and Weibull are available in the reliability literature.
Series systems. As for the static case, the reliability of a series system (Figure 19.1) is the product of the component reliabilities,

$$R = \prod_{i=1}^{m} R_i$$

The failure density function,

$$f = -\frac{dR}{dt}$$

$$= -\frac{dR_1}{dt} R_2 \ldots R_m - R_1 \frac{dR_2}{dt} R_3 \ldots R_m - \ldots - R_1 R_2 R_3 \ldots \frac{dR_m}{dt}$$

and the failure rate,

$$\lambda = \frac{f}{R}$$

$$= -\frac{dR_1 / dt}{R_1} - \frac{dR_2 / dt}{R_2} - \ldots - \frac{dR_m / dt}{R_m}$$

$$= \frac{f_1}{R_1} + \frac{f_2}{R_2} + \ldots + \frac{f_m}{R_m}$$

$$= \lambda_1 + \lambda_2 + \ldots + \lambda_m$$

That is, the failure rate of the system is the sum of the failure rates of the components.

Parallel systems. Parallel systems are also called redundant systems and include standby systems. Some particular types of parallel systems include:

- Hot redundancy. All components are capable of being used, and are available if still functioning. Failure of the system requires failure of all components. An example is a twin-engine aircraft.
- Cold or standby redundancy. Components are only brought into operation when needed, as other parallel components fail. Components lay dormant until needed and hence they age less compared to components in a hot redundancy, but do require activation by a switch-type mechanism to bring into use. An example is the spare wheel-tyre in a car.

For the hot redundancy system (Figure 19.3), failure occurs when all components fail.

$$F = F_1 \times F_2 \times \ldots \times F_m$$

or

$$R = 1 - (1 - R_1)(1 - R_2) \ldots (1 - R_m)$$

Unlike the series case, there is no straightforward general relationship between the system failure rate and the component failure rates. Each failure rate assumption has to be worked through on its own.

19.5 STRUCTURAL ENGINEERING

Of all the disciplines, structural engineering comes closest to the correct understanding of risk. This is possibly because of the extensive work over many years in reliability analysis and limit states design, and having good system models.

Chapter 20

Fault trees, event trees

20.1 INTRODUCTION

Analysis, within risk management, uses conventional methods such as those used elsewhere including forecasting, sensitivity, Monte Carlo simulation and fault and event trees. Such techniques existed before the discipline and formalism of risk management became popular. Fault tree and event tree methods are borrowed from the area of reliability. All analysis methods borrowed by risk management together might be referred to as 'risk analysis', 'quantitative risk assessment' or 'probabilistic risk assessment', but all such terms are misleading. In essence, analysis is the conversion of inputs (risk sources) into outputs (or in reverse for fault trees). Evaluation is done subsequent to analysis.

Both fault trees and event trees may be used to provide information for the evaluation step of risk management. Fault trees and event trees, besides as forms of analysis, can also assist in risk source identification.

20.2 OVERVIEW

Fault trees address the question: How can failure occur? What could happen that would give rise to or cause a failure? Fault tree analysis starts from a (root) fault and looks at what might cause the fault, and then what might cause these causes and so on. Event trees answer the question: What could happen or follow if a given event occurs? Event tree analysis traces subsequent events leading from an initiating event.

The terminology used here for fault trees and event trees is terminology favoured in the fault tree and event tree literature, rather than the more generic terminology preferred in the body of this book. The two sets of terminology are related as follows. A fault tree goes from a top event (failure, loss, accident) through contributing events to primary events; in the terminology of this book, the input or risk source are the primary events, while the output or consequence is the top event. An event tree goes from an

DOI: 10.1201/9781003343592-22

initiating event to subsequent events to consequences; in the terminology of this book, the input or risk source is the initiating event, while the consequences are the outputs.

Fault trees and event trees provide a systematic way of identifying faults, losses, unwanted events, failures, ..., their contributing causes and consequences. Each may provide insight through their systematic and logical development. They are diagnostic tools given strength by the pictorial way they represent the information.

The starting point for a fault tree is usually a system failure (the top event) and the causes are traced back through a collection of AND/OR gates to the possible initiating sources.

An event tree starts with an input (initiating fault or event) and works forward (in time) through a sequence of ensuing failures or events to outputs (consequences) and their probabilities. For any given hazard or failure situation, risk sources (inputs) are identified, possible sequences that could follow from these are traced through event trees, probabilities are attached and probabilities are calculated for outputs (consequences).

Event trees and fault trees might be considered as <u>bottom-up</u> or <u>top-down</u> approaches, respectively. Each tree type relates outputs (failures or consequences) to component sources. If past data are rare, to predict the probability of failure of a system is difficult. However, by looking at the components and possible chains that could lead to failure, then some idea of the probability of failure of the system may be obtained. Similar comments apply to working out the type of failure.

Fault trees and event trees can be related. In particular, the probability of an event in an event tree can be evaluated using a fault tree drawn with that event as the top event.

Fault trees and event trees might be criticized for their assumed independence of events, for neglecting common mode failures, and for relying on data that may not exist but rather have to be estimated. However, the two tree types are useful in helping to understand a failure-type situation.

20.3 FAULT TREES

Outline. A fault tree offers a systematic and organized way of unearthing possible and potential faults and their effects on a failure event. Probabilities of the various faults or failure events may be incorporated into the overall analysis, but this is not necessary. The analysis is a useful diagnostic tool for identifying failure paths and critical events.

Fault trees can be used to identify failure modes and sequences of events leading to failure. Human nature tends to want to eliminate failure modes and failure sequences or reduce their probability of occurrence. In such an approach, the probability associated with the head event is of lesser importance than gaining an understanding of the failure.

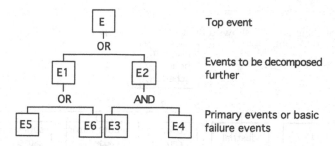

Figure 20.1 Example fault tree.

In a fault tree diagram, the failure event, 'top event' or 'head event' is decomposed into subevents (faults) which can be still further decomposed until 'basic events' are reached. Typically, probabilities can be associated with basic events.

The top or head event is the loss, accident or unwanted event. Subevents are contributors to this. The links in the fault tree identify the sequences that lead to the top event. The tree permits the thinking through of the possible causes (inputs) of a loss (output) and quantifies the probability of the loss.

Depending on the relationship between events, so their union or intersection is considered. This is represented by logic gates – 'OR' gates (for union) and 'AND' gates (for intersection) (for example, Figure 20.1). In Figure 20.1, either event E_5 or E_6 could lead to event E_1, while both events E_3 and E_4 are needed to lead to event E_2.

Some literature use different geometric shapes for different types of events and use symbols for the gates. Use of such symbolism is optional.

Example: In the collapse of a building, the potential modes of failure and their respective causes may be represented by the fault tree shown in Figure 20.2.

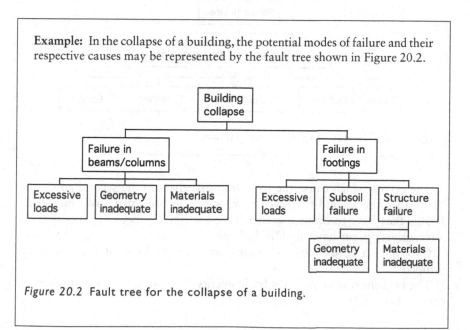

Figure 20.2 Fault tree for the collapse of a building.

Example: Figure 20.3 shows a fault tree for a project behind schedule.

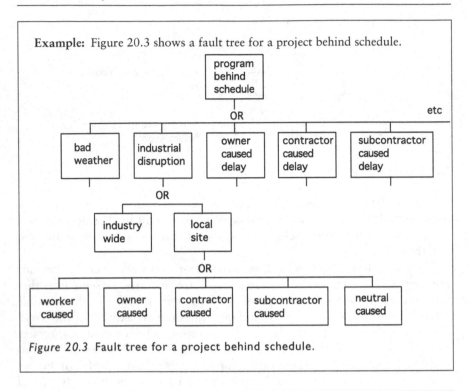

Figure 20.3 Fault tree for a project behind schedule.

Example: Consider the failure of brakes on a vehicle (Figure 20.4).

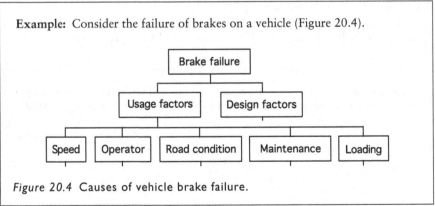

Figure 20.4 Causes of vehicle brake failure.

The events which cause the top event are given in the second row, those that cause the second-row events are given in the third row, and so on. Events that have to occur together have an 'AND' linkage; events that can alternatively occur have an 'OR' linkage.

Care has to be exercised so that a step in the logic of the tree is not jumped.

The bottom row constitutes basic events that cannot be decomposed, or there is no need to decompose, further.

The fault tree develops from the head event down. The choice of this event is therefore important. It may be a failure event or loss resulting from that failure. Each will give a different fault tree.

Example: Consider a contractual matter (Figure 20.5).

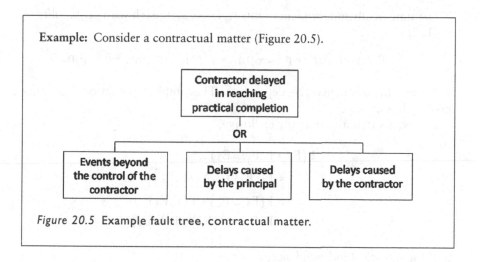

Figure 20.5 Example fault tree, contractual matter.

Probability evaluation. The probability of the top event occurring can be related through the combinations of the 'AND' and 'OR' gates to the probabilities of the subevents occurring. 'AND' is interpreted as an intersection of events – for independent events the probabilities are multiplied. 'OR' is interpreted as a union of events – the probabilities are added.

Using Figure 20.1 as an example, and starting at the bottom of the tree and working upwards,

$$P(E_1) = P(E_5 \text{ or } E_6)$$

$$P(E_2) = P(E_3 \text{ and } E_4)$$

and

$$P(E) = P(E_1 \text{ or } E_2)$$

In more formal notation,

$$
\begin{aligned}
P(E) &= P(E_1 \cup E_2) \\
&= P\left[(E_5 \cup E_6) \cup E_3 E_4\right] \\
&= P(E_5 \cup E_6) + P(E_3 E_4) - P\left[E_3 E_4(E_5 \cup E_6)\right] \\
&= P(E_5) + P(E_6) - P(E_5 E_6) + P(E_3 E_4) \\
&\quad - P(E_3 E_4 E_5) - P(E_3 E_4 E_6) + P(E_3 E_4 E_5 E_6)
\end{aligned}
$$

where,

P(E) the probability of the top event
$P(E_i)$ probability of event E_i

For statistically independent events E_i, i = 1, 2, ... each with probability p_i, i = 1, 2, ...

$$P(E) = p_5 + p_6 - p_5 p_6 + p_3 p_4 - p_3 p_4 p_5 - p_3 p_4 p_6 + p_3 p_4 p_5 p_6$$

Events, however, may be dependent. This implies conditional or joint probabilities.

For events that are mutually exclusive,

$$
\begin{aligned}
P(E) &= P(E_1 \cup E_2) \\
&= P\left[(E_5 \cup E_6) \cup E_3 E_4 \right] \\
&= P(E_5 \cup E_6) + P(E_3 E_4) \\
&= P(E_5) + P(E_6) + P(E_3 E_4)
\end{aligned}
$$

and for statistical independence,

$$P(E) = p_5 + p_6 + p_3 p_4$$

Other aspects. The emphasis above was on the quantitative evaluation of fault trees, relying on the establishment of probabilities for the basic events. However, the development of fault trees, in itself, is a good discipline for examining causes of an accident or failure, and leads to a systematic breakdown of causes. It can be used to think through the causes of a loss, failure or accident, and to investigate failures.

The tree may also be examined qualitatively giving indicative results.

Dominant probabilities within the tree may be identified and the associated branches carried through the analysis at the expense of events with small probabilities. This helps prune large fault trees to a manageable size.

Having calculated the probability associated with the top event, this feeds into any subsequent evaluation. The *Response* step may adjust something in the tree. Typically, something or some data within the tree are altered to see its effect on reducing the probability of the head event. Alternatively, a trade-off with cost of changing something or some data may be of interest.

Sensitivity. Given that the accuracy in the probability estimates may be lacking, a sensitivity analysis is recommended. Analyses need to be carried out over the whole range of event probabilities.

The end result may be that the head event probability can only be established to an order of magnitude or that the head event probability ranges over several orders of magnitude.

Responses can then be made in this light.

Probability estimates. Probabilities of events may be difficult to obtain in some cases, particularly those events relating to human error and to events that are rare. In many cases, best guesses are all that are available. Probabilities associated with equipment, component and like failures may be obtained through historical or test data. Those associated with financial trends and legal matters may be obtained by seeking the opinions of a number of people.

20.4 EVENT TREES

Outline. Event tree diagrams permit the tracing of the outputs (consequences) resulting from an input (initiating event) and subsequent events (Figure 20.6). The top event is the input (initiating event) and the results of this event are established through following the sequences of subsequent events. Particular outputs (consequences) can be traced through paths within the tree.

The input (initiating event) may be a failure, unwanted event or similar. Event trees are commonly applied to an accident scenario. Developing an event tree enables all potentially dangerous or adverse outputs (consequences) to be considered in a systematic fashion.

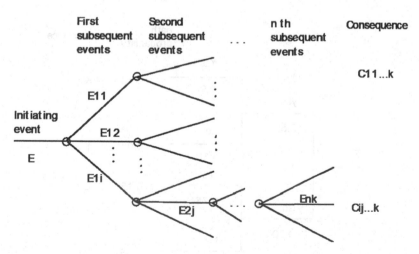

Figure 20.6 General event tree (based on Ang and Tang, 1984).

The general event tree of Figure 20.6 has outputs (consequences) $C_{ij...k}$. Each path through the tree represents a specific sequence of (subsequent) events.

The probability associated with the occurrence of any particular output (consequence) is the product of the (conditional) probabilities of all the events on the associated path.

$$P\left(C_{ij...k}\,|\,E\right) = P\left(E_{1i}\,|\,E\right)P\left(E_{2j}\,|\,E_{1i}\,E\right)...P\left(E_{nk}\,|\,E_{1i}\,E_{2j}...E\right)$$

where E is the initiating event.

Calculations. Figure 20.7 is an example to show the calculations involved with event trees. The probabilities associated with different outputs (consequences) are evaluated through the relevant paths of the event tree.

$$P\left(Y\,|\,E\right) = 0.80 \times 0.10 + 0.20 \times 0.25 = 0.13$$

$$P\left(Z\,|\,E\right) = 0.20 \times 0.75 = 0.15$$

$$P\left(X\,|\,E\right) = 0.80 \times 0.95 = 0.72$$

Note, the sum of the probabilities of all outputs (consequences) is 1. Also, at each stage the probabilities of the yes and no options sum to 1; either the event will happen or it will not happen – each is the complement of the other.

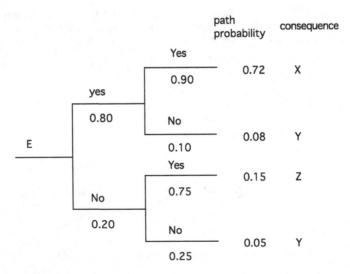

Figure 20.7 Example event tree.

Event trees may be used qualitatively to identify outputs (consequences) and sequences of events leading from an input (initiating event) to an output (consequence). They do not have to be used quantitatively, particularly if the values of the event probabilities are difficult to obtain or estimate. They are a useful tool for developing logic in a systematic fashion.

Example: Figures 20.8 and 20.9 give two examples of an event tree resulting from the finding of a latent site condition in the performance of work under a contract.

Common events resulting from the finding of a latent site condition are:

- The contract will be frustrated.
- A variation to the work as specified [the permanent works] will be necessary.

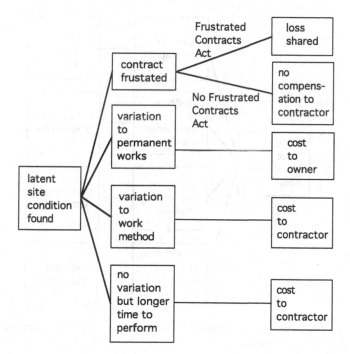

Figure 20.8 Event tree for latent site condition – no latent conditions clause in contract.

- A variation to the contractor's proposed method of working [the temporary works] will be necessary.
- No variation is necessary but the work will take longer to perform.

Figure 20.8 corresponds to the situation for the contract where there exists no latent site condition clause and hence common law applies. Where legislation exists covering frustration, the loss could be anticipated to be shared between the owner and the contractor. Where there is no such legislation, the contractor may not be entitled to any compensation under the contract, although a restitution claim may exist.

Figure 20.9 applies to a contract that allows for the occurrence of latent conditions. Generally, the owner bears any risk associated with a latent site condition. However, the contractor may still bear some exposure such as loss of profit in respect of delay or disruption.

Further examples of event trees based on contracts are given in Figures 20.10 and 20.11.

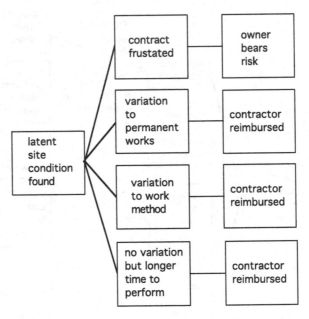

Figure 20.9 Event tree for latent site condition – latent conditions clause in contract.

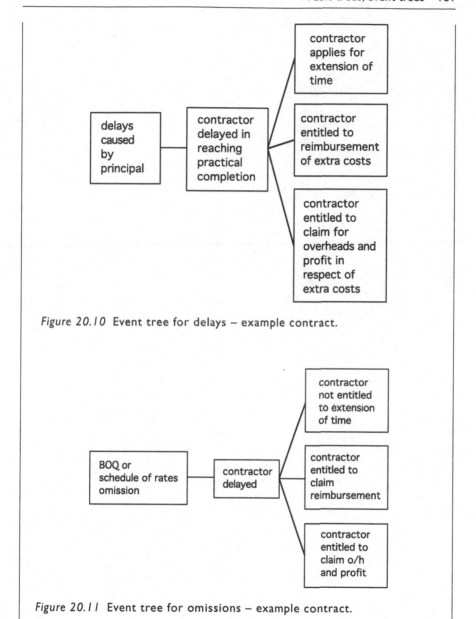

Figure 20.10 Event tree for delays – example contract.

Figure 20.11 Event tree for omissions – example contract.

References

Al-Sadek O; Carmichael DG, 1992. On simulation in planning networks, *Civil Engineering and Environmental Systems*, vol. 9, pp. 59–68.

Ang AH-S; Tang WH, 1975. *Probability Concepts in Engineering Planning and Design, Vol. I*, Wiley, New York.

Ang AH-S; Tang HW, 1984. *Probability Concepts in Engineering Planning and Design, Vol. II*, John Wiley and Sons, New York.

Antill JM; Woodhead RW, 1970. *Critical Path Methods in Construction Practice*, 2nd edition, Wiley-Interscience, New York.

AS/NZS, 1995. *AS/NZS 4360 Risk Management*, Standards Australia, Sydney; Standards New Zealand, Wellington.

Balatbat MC; Findlay E; Carmichael DG, 2012. Performance risk associated with renewable energy CDM projects, *Journal of Management in Engineering*, vol. 28, pp. 51–58, http://dx.doi.org/10.1061/(ASCE)ME.1943-5479.0000090

Benjamin JR; Cornell CA, 1970. *Probability, Statistics and Decision for Civil Engineers*, McGraw-Hill, New York.

Carmichael DG, 1981. *Structural Modelling and Optimization*, Ellis Horwood Ltd (John Wiley and Sons), Chichester, UK.

Carmichael DG, 1987. *Engineering Queues in Construction and Mining*, Ellis Horwood Ltd (John Wiley and Sons), Chichester, UK.

Carmichael DG, 1988. Queue simulation of cyclic construction operations, *Civil Engineering Systems*, vol. 5, pp. 213–219.

Carmichael DG, 1989a. *Construction Engineering Networks*, Ellis Horwood Ltd (John Wiley and Sons), Chichester, UK.

Carmichael DG, 1989b. Production tables for earthmoving, quarrying and open-cut mining, in Carmichael DG (ed.), *Applied Construction Management*, Unisearch, The University of New South Wales, Sydney, pp. 275–284.

Carmichael DG, 1990. Structural optimization and system dynamics, *Structural Optimization*, vol. 2, pp. 105–108.

Carmichael DG, 1996. Flat organisational structures, *Journal of Project and Construction Management*, vol. 2, pp. 61–68.

Carmichael DG, 1997a. Management fads, *Journal of Project and Construction Management*, vol. 3. pp. 115–125.

Carmichael DG, 1997b. Re-engineering and work study, *Journal of Project and Construction Management*, vol. 3, pp. 95–107.

Carmichael DG, 1998. Gurus of faddish management, *Journal of Project and Construction Management*, vol. 4, pp. 77–84.

Carmichael DG, 2000. *Contracts and International Project Management*, A.A. Balkema, Rotterdam, The Netherlands.

Carmichael DG, 2002. *Disputes and International Projects*, A.A. Balkema Publishers, Rotterdam, The Netherlands.

Carmichael DG, 2004. *Project Management Framework*, A.A. Balkema Publishers, Lisse, The Netherlands.

Carmichael DG, 2006. *Project Planning, and Control*, Taylor & Francis, Oxon, UK

Carmichael DG, 2009. Comments on delay analysis methods in resolving construction claims, *International Journal of Construction Management*, vol. 9, pp. 1–28.

Carmichael DG, 2011. An alternative approach to capital investment appraisal, *Engineering Economist*, vol. 56, pp. 123–139.

Carmichael DG, 2013a. *Problem Solving for Engineers*, CRC Press, Taylor and Francis, London.

Carmichael DG, 2013b. The conceptual power of control systems theory in engineering practice, *Civil Engineering and Environmental Systems*, vol. 30, pp. 231–242.

Carmichael DG, 2014. *Infrastructure Investment: An Engineering Perspective*, CRC Press, Taylor and Francis, London.

Carmichael DG, 2015. Incorporating resilience through adaptability and flexibility, *Civil Engineering and Environmental Systems*, vol. 32, pp. 31–43.

Carmichael DG, 2016a. Risk–a commentary, *Civil Engineering and Environmental Systems*, vol. 33, pp. 177–198.

Carmichael DG, 2016b. A cash flow view of real options, *Engineering Economist*, vol. 61, pp. 265–288.

Carmichael DG, 2017. Adjustments within discount rates to cater for uncertainty— Guidelines, *Engineering Economist*, vol. 62, pp. 322–335.

Carmichael DG, 2019. Organisations as systems - difficulties in model development and validation, *Civil Engineering and Environmental Systems*, vol. 35, pp. 41–56.

Carmichael DG, 2020a. *Future-proofing–Valuing Adaptability, Flexibility, Convertibility, and Options; A Cross-disciplinary Approach*, Springer Publishers, Berlin, 2020.

Carmichael DG, 2020b. Bias and decision making - An overview systems explanation, *Civil Engineering and Environmental Systems*, vol. 37, nos 1–2, pp. 48–61.

Carmichael DG, 2020c. A framework for a civil engineering systems BOK, *Civil Engineering and Environmental Systems*, vol. 37, no. 4, pp. 154–165.

Carmichael DG, 2020d. Author's reply to: David Elms' discussion of 'Bias and decision making - An overview explanation', *Civil Engineering and Environmental Systems*, vol. 37, no. 3, pp. 146–148.

Carmichael DG, 2021a. At one with systems (Special issue - Civil engineering systems body of knowledge, discussion paper), *Civil Engineering and Environmental Systems*, vol. 38, no. 4, pp. 265–268.

Carmichael DG, 2021b. BOK and terminology (Special issue - Civil engineering systems body of knowledge, discussion paper), *Civil Engineering and Environmental Systems*, vol. 38, no. 4, pp. 257–258.

Carmichael DG, 2021c. Author's reply to: David Elms' discussion of 'A framework for a civil engineering systems BOK', *Civil Engineering and Environmental Systems*, vol. 38, no. 4, pp. 276–278.

Carmichael DG, 2021d. Author's reply to: David Blockley's discussion of the special Issue, *Civil Engineering and Environmental Systems*, vol. 38, no. 4, p. 250.

Carmichael DG; Balatbat MC, 2008. Probabilistic DCF analysis and capital budgeting and investment - A Survey, *The Engineering Economist*, vol. 53, pp. 84–102.

Carmichael DG; Balatbat MCA, 2010. A contractor's analysis of the likelihood of payment of claims, *Journal of Financial Management of Property and Construction*, vol. 15, pp. 102–117.

Carmichael DG; Balatbat MCA, 2011a. On the analysis of property unit sales over time, *International Journal of Strategic Property Management*, vol. 15, pp. 329–339.

Carmichael DG; Balatbat MCA, 2011b. Risk associated with managed investment primary production projects, *International Journal of Project Organisation and Management*, vol. 3, pp. 273–289.

Carmichael DG; Ballouz JJ; Balatbat MCA, 2015. Improving the attractiveness of CDM projects through allowing and incorporating options, *Energy Policy*, vol. 86, pp. 784–791.

Carmichael DG; Bartlett BJ; Kaboli AS, 2014. Surface mining operations: Coincident unit cost and emissions, *International Journal of Mining, Reclamation and Environment*, vol. 28, pp. 47–65.

Carmichael DG; Bustamante BL, 2014. Interest rate uncertainty and investment value: A second order moment approach, *International Journal of Engineering Management and Economics*, vol. 4, pp. 176–176.

Carmichael DG; Edmondson CG, 2015. Risk in stream and royalty financing of infrastructure development, *Journal of Infrastructure Development*, vol. 1, pp. 23–40.

Carmichael DG; Handford LB, 2015. A note on equivalent fixed rate and variable rate loans; borrower's perspective, *The Engineering Economist*, vol. 60, pp. 155–162.

Carmichael DG; Hersh AM; Parasu P, 2011. Real options estimate using probabilistic present worth analysis, *The Engineering Economist*, vol. 56, pp. 295–320.

Carmichael DG; Karantonis JP, 2015. Construction contracts with conversion capability: A way forward, *Journal of Financial Management of Property and Construction*, vol. 20, pp. 132–146.

Carmichael DG; Lea KA; Balatbat MCA, 2016. The financial additionality and viability of CDM projects allowing for uncertainty, *Environment, Development and Sustainability*, vol. 18, pp. 129–141.

Carmichael DG; Mustaffa NK, 2018. Emissions and production penalties/bonuses associated with non-standard earthmoving loading policies, *Construction Innovation*, vol. 18, pp. 227–245.

Carmichael DG; Mustaffa NK; Shen X, 2018. A utility measure of attitudes to lower-emissions production in construction, *Journal of Cleaner Production*, vol. 202, pp. 23–32.

Carmichael DG; Nguyen TA; Shen X, 2019a. Single treatment of PPP road project options, *Journal of Construction Engineering and Management*, vol. 145, pp. 04018122-1–04018122-11.

Carmichael DG; Shen X; Peansupap V, 2019b. The relationship between heavy equipment cost efficiency and cleaner production in construction, *Journal of Cleaner Production*, vol. 211, pp. 521.

Hall AD, 1962. *A Methodology for Problem Solving*, D. Van Nostrand, Princeton.

Hosseinian SM; Carmichael DG, 2013. Optimal gainshare/painshare in alliance projects, *Journal of the Operational Research Society*, vol. 64, pp. 1269–1278.

Hosseinian SM; Carmichael DG, 2014a. Optimal sharing arrangement for multiple project outcomes, *Journal of Financial Management of Property and Construction*, vol. 19, pp. 264–280.

Hosseinian SM; Carmichael DG, 2014b. Optimum sharing in project delivery methods, *International Journal of Engineering Management and Economics*, vol. 4, pp. 151–175.

Hosseinian SM; Carmichael DG, 2016. Optimal outcome sharing with a consortium of contractors, *Journal of Civil Engineering and Management*, vol. 22, pp. 655–665.

Hosseinian SM; Farahpour E; Carmichael DG, 2020. Optimum outcome-sharing construction contracts with multi-agent and multi-outcome arrangements, *Journal of Construction Engineering and Management*, vol. 146, no. 7, 12 pages.

ISO, 2009. *ISO 31000 Risk Management*, International Organization for Standardization, Geneva Switzerland.

Jukic D; Carmichael DG, 2016. Emission and cost effects of training for construction equipment operators: A field study, *Smart and Sustainable Built Environment*, vol. 5, pp. 96–110,

Lehmann DR, 1979. *Market Research and Analysis*, Irwin, Homewood Illinois USA.

Parkinson CN, 1958. *Parkinson's Law or the Pursuit of Progress*, John Murray, London.

Parkinson CN, 1960. *The Law and the Profits*, John Murray, London.

Safe Work Australia, 2018. *Construction Work Code of Practice*, Safe Work Australia, Canberra, May.

Tan F; Makwasha T, 2010. Best practice cost estimation in land transport infrastructure projects, *Australasian Transport Research Forum*, Canberra Australia, 29 September–1 October, 'Best practice' cost estimation in land transport infrastructure projects [Accessed 11 May 2020].

Tran H; Carmichael DG, 2012a. Contractor's financial estimation based on owner payment histories, *Organization, Technology and Management in Construction: An International Journal*, vol. 4, no. 2, pp. 481–489.

Tran H; Carmichael DG, 2012b. The likelihood of subcontractor payment; downstream progression via the owner and contractor, *Journal of Financial Management of Property and Construction*, vol. 17, pp. 135–152.

Tran H; Carmichael DG, 2013. A contractor's classification of owner payment practices, *Engineering, Construction and Architectural Management*, vol. 20, pp. 29–45.

Tribus M, 1969. *Rational Descriptions, Decisions, and Designs*, Pergamon Publishing Company, Elmsford, New York.

Wood DA, 2002. Risk simulation techniques to aid project cost-time planning and management, *Risk Management*, vol. 4, pp. 41–60.

Index

Printed in the United States
by Baker & Taylor Publisher Services